Nucleic Acids and Molecular Biology

17

Series Editor
H. J. Gross

Wolfgang Nellen   Christian Hammann   (Eds.)

Small RNAs:
Analysis and Regulatory Functions

With 32 Figures, and 6 Tables

 Springer

Professsor Dr. Wolfgang Nellen
Dr. Christian Hammann
Dept. of Genetics
University of Kassel
Heinrich-Plett-Str. 40
34132 Kassel
Germany

ISSN 0933-1891
ISBN-10 3-540-28129-0   Springer Berlin Heidelberg New York
ISBN-13 978-3-540-28129-0   Springer Berlin Heidelberg New York

Library of Congress Control Number: 2005934658

Springer is a part of Springer Science+Business Media

springeronline.com

© Springer-Verlag Berlin Heidelberg 2006
Printed in Germany

Production and typesetting: LE-TEX Jelonek, Schmidt & Vöckler GbR, 04229 Leipzig
Cover design: *design & production* GmbH, 69126 Heidelberg

Printed on acid-free paper   31/3150/YL - 5 4 3 2 1 0

*The authors dedicate this volume
to the memory of Martin Tabler,
a friend and highly respected colleague
who died in a tragic accident on April 3rd, 2005.*

Preface

When microRNAs were first described in 1993 by the groups of Ruvkun (Wightman et al., 1993) and Ambros (Lee et al., 1993) or possibly even in 1976 by Heywood (Heywood and Kennedy, 1976), the scientific community was not yet ready to accept RNA as a general regulator of gene expression. Small non-coding prokaryotic RNAs were among the first to draw attention to their unusual structures and their potential to react rapidly to environmental stimuli and cellular changes by regulating genes on the post-transcriptional level (Mizuno et al., 1984). It is therefore appropriate to begin this volume with an update by Wagner and Darfeuille on small, non-coding RNAs in prokaryotes, that control cellular processes on many different levels.

More specifically, Söderbom describes approaches to identify regulatory RNAs *in silico* and *in vivo* and concentrates on snoRNAs. Though these are known to serve as guides for modification of rRNA and snRNAs, further targets are emerging.

The *in silico* search for non-coding RNAs is detailed in the contribution of Gräf et al. who describe programs to find non-coding RNAs in large databases.

Boutla and Tabler discuss the dual approach of bioinformatics and *in vivo* experiments to define and validate new microRNAs – a difficult task of great importance to understand this formerly unexpected network of regulation in eukaryotic development and differentiation.

Secondary structure, in particular dsRNA, has become a major focus in gene regulation. DsRNA is a processing intermediate and recruits diverse machineries that modify RNA, elicit enzymatic responses or guide RNA to specific targets. In this context, the proteins that interact with dsRNA have attracted increasing attention. The chapter by Hammann reviews the features of dsRBDs and discusses the problem how target dsRNAs are guided to one but not the other molecular pathway that coexist in the cell.

The following two chapters by Bleys et al. and Kuhlmann et al. concentrate on special aspects of the RNA interference pathway. In contrast to the basic RNAi mechanism, the spreading of dsRNA mediated gene silencing in an organism (systemic silencing) and along the gene or genome (transitive silencing) most likely involves the activity of RNA directed RNA polymerases. Models on how these enzymes may be involved in signal generation and amplification are also presented in the latter chapter that evaluates differences and similarities between RNAi and antisense mediated gene silencing.

One branch of the RNA interference mechanism acts in the nucleus and confers transcriptional silencing by influencing chromatin structure. Dijk et al. review experiments that explain how small interfering RNAs prevent the proliferation of transposons in the genome of the model organism Chlamydomonas. Paulsen and co-workers extend the view on epigenetic regulation and summarize examples of chromatin remodelling at specific imprinted genes and in dosage compensation in the general context of RNA interference. Finally, Sano and Taira describe the application of small artificial riboregulators to identify gene functions in mammalian cells.

This volume can only provide a snapshot of the rapidly developing field of small RNA regulatory functions. New details are continuously emerging, nevertheless, we believe that the overviews, speculations, methods and models presented in this book will not only provide a summary of the current knowledge but also stimulate discussions and ideas how to further tackle this exciting area of research.

Kassel, June 2005   *Wolfgang Nellen*
                                                                       Christian Hammann

References

Heywood SM, Kennedy DS (1976) Purification of myosin translational control RNA and its interaction with myosin messenger RNA. Biochemistry 15:3314–3319

Lee RC, Feinbaum RL, Ambros V (1993) The C. elegans heterochronic gene lin-4 encodes small RNAs with antisense complementarity to lin-14. Cell 75:843–854

Mizuno T, Chou MY, Inouye M (1984) A unique mechanism regulating gene expression: translational inhibition by a complementary RNA transcript (micRNA). Proc Natl Acad Sci USA 81:1966–1970

Wightman B, Ha I, Ruvkun G (1993) Posttranscriptional regulation of the heterochronic gene lin-14 by lin-4 mediates temporal pattern formation in C. elegans. Cell 75:855–862

Contents

Transitive and Systemic RNA Silencing:
Both Involving an RNA Amplification Mechanism?

RNA Interference and Antisense Mediated Gene Silencing

Nucleic Acids and Molecular Biology, Vol. 17
Wolfgang Nellen, Christian Hammann (Eds.)
Small RNAs
© Springer-Verlag Berlin Heidelberg 2005

Small Regulatory RNAs in Bacteria

E. Gerhart H. Wagner (✉) · Fabien Darfeuille

Department of Cell and Molecular Biology, Uppsala University, Box 596, 75124 Uppsala,
Sweden
gerhart.wagner@icm.uu.se

Abstract In recent years, small regulatory RNAs have been discovered at a staggering rate
both in prokaryotes and eukaryotes. By now it is clear that post-transcriptional regulation of gene expression mediated by such RNAs is the rule rather than—as previously
believed—the exception. In this chapter, we focus on small RNAs (sRNAs) encoded by
bacterial chromosomes. The strategies for their discovery, their biological roles, and their
mechanisms of action are discussed. Even though the number of well-characterized sR-
NAs in, for example, the best studied model enterobacterium *Escherichia coli*, is still
small, the emerging pattern suggests that antisense mechanisms predominate. In terms
of their roles in bacterial physiology, most of these RNAs appear to be involved in stress
response regulation. Some other examples indicate functions in regulation of virulence.
Two aspects of sRNA-mediated control arising from recent observations are addressed
as well. Firstly, some sRNAs need proteins (notably Hfq) as helpers in their antisense
activities—at this point the reason for this requirement is not understood. Secondly, only
limited sequence complementarity is generally observed in antisense–target RNA pairs.
This raises the fundamental question of how specific recognition is accomplished, and
what the structure/sequence determinants for rapid and productive interaction are.

1
Introduction

Textbooks tell us that regulation of gene expression, in all organisms, is almost exclusively a task for regulatory proteins—activators and repressors. To
a great extent, this view draws from a focus on transcriptional control; additional levels of control are often considered to be of lesser importance. However,
as more and more cases of post-transcriptional control are being uncovered,
other players enter the stage. Already in the early 1980s, a class of RNAs, denoted
antisense RNAs, were shown to act as key regulators of plasmid replication
(Stougaard et al. 1981; Tomizawa et al. 1981). Subsequently, many other antisense RNAs were found to control plasmid copy number, postsegregational
killing and transposition in bacteria, and to regulate lysis/lysogeny decisions in
bacteriophages. Thus, in line with an early (but often forgotten) hypothesis by
Jacob and Monod (1961), regulation of gene expression in bacteria can indeed
be carried out by RNA—as well as by proteins.

The discovery of noncoding, antisense RNAs as regulators of gene expression in accessory genetic elements (Wagner et al. 2002) and their exquisite

properties—with respect to efficiency and specificity—suggested the possibility that chromosomally encoded genes might also make use of this general strategy; however, up until recently, only a handful of such regulatory RNAs had been found in bacteria, and even fewer in eukaryotes. This picture changed dramatically in 2001, when several genome-wide searches were conducted in the model bacterium *Escherichia coli*, uncovering more than 30 new small RNAs (sRNAs) encoded by intergenic regions (IGRs), or more precisely inter-open reading frame (ORF) regions (Argaman et al. 2001; Rivas et al. 2001; Wassarman et al. 2001). Additional searches (see later) brought this number to about 70 RNAs that we know of today. Thus, many previously undetected sRNAs are present—many of which are conserved between bacterial relatives—but what are their biological roles?

This chapter will review our current understanding of the functions these sRNAs appear to carry out in bacteria, and address mechanistic aspects of regulation—though it is clear that this field is still in its infancy. The strong focus on sRNAs from *E. coli* simply reflects that, at this point in time, only very few regulatory RNAs have been identified in other bacteria. Some of these will be addressed here. In the last few years, many reviews have addressed the prevalence of these new sRNAs, the strategies for their discovery and their functional characterization (Altuvia and Wagner 2000; Gottesman 2004; Gottesman et al. 2001; Storz et al. 2004; Vogel and Wagner 2005; Wagner and Flärdh 2002; Wagner and Vogel 2003, 2005; Wassarman 2002).

As far as regulatory RNAs in eukaryotic organisms are concerned—a topic covered elsewhere in this book—a similar picture has emerged. In the aftermath of the discovery of RNA interference (He and Hannon 2004), the size similarity (approximately 22 nucleotides in length) between small interfering RNAs (siRNAs) and Lin-4 and Let-7, two known sRNAs that are known to regulate developmental genes in the worm *Ceanorhabditis elegans*, suggested shared pathways of their generation. This prompted several groups to initiate searches for sRNAs. Today we know that hundreds of RNAs of this size are encoded by metazoan and plant chromosomes. These RNAs, denoted microRNAs (miRNAs), act as antisense RNAs to target, it seems, primarily the messenger RNAs (mRNAs) of developmental genes (Bartel 2004).

2
Searches and Discovery of sRNAs

Genes encoding sRNAs are easy to overlook. Unless a lucky accident gives a phenotype in a mutant screen, such genes tend to remain unknown. In addition, on the basis of the last few years of work, it is apparent that mutations in sRNA genes usually confer, at best, subtle phenotypes. The availability of genomic sequences initially provided no remedy; protein-encoding genes could most often be identified by bioinformatics, whereas sRNA-encoding genes escaped detection and were not annotated. Thus, in the past, discovery of noncoding RNAs in bacteria (and elsewhere) has been mostly serendipitous.

An additional obstacle to the realization that sRNAs are in fact prevalent could be labeled prejudice. The scientific community by and large did not recognize sRNAs as major players in, for example, control of physiology or development. On top of that, the dense packing of genes in bacterial chromosomes, with only short empty spaces (IGRs) between protein-encoding genes, reinforced expectations that few, if any, sRNA-encoding genes might be present.

In line with this, the total list of noncoding RNAs identified over a 30-year period was short: apart from the housekeeping RNAs (transfer RNAs, tRNAs, ribosomal RNAs, the RNA subunit of RNase P, tmRNA, and 4.5S—the RNA subunit of the signal recognition particle), only a few sRNAs had been found by the end of the 1990s (Wassarman et al. 1999). Housekeeping RNAs will not be discussed in this review.

2.1
The Early Years

The first sRNAs were identified after fractionation of in vivo ^{32}P-labeled total bacterial RNA. Abundant RNAs could be isolated on gels, autoradiographed, and fingerprinted. This gave, for example, 6S RNA and Spot 42 (Hindley 1967; Ikemura and Dahlberg 1973)—both remained functionally undefined until recently.

A few sRNAs were discovered in genetic screens that were designed to identify trans-acting factors that modulated/controlled certain genes. For example, the introduction of multiple copies of a cloned genomic DNA segment resulted in depletion of the OmpF (outer membrane porin) protein in E. coli cells. The insert was shown to encode MicF, the first chromosomally encoded antisense RNA, which base-pairs to the leader of the ompF mRNA entailing translational repression (Andersen et al. 1989). In a similar fashion, a regulator of hns (encoding a histone-like protein/global transcriptional regulator) was found in a screen for genes affecting capsule synthesis (Sledjeski and Gottesman 1995). This regulator, DsrA, is an antisense RNA which targets the hns mRNA and inhibits its translation. Subsequent work showed that, remarkably, DsrA has dual functions. Its second target, the rpoS (stress Sigma factor S, σ^S) mRNA, is activated. This occurs by a folding change—caused by DsrA binding—of the mRNA leader such that the translation initiation region (TIR) becomes accessible to ribosomes (Majdalani et al. 1998). We will return to mechanistic questions later. In a second genetic screen, another sRNA, RprA, was found to target (and upregulate) rpoS as well (Majdalani et al. 2001). The emerging theme of more than one sRNA targeting the same mRNA, or a single sRNA targeting several mRNAs, may have interesting physiological consequences. The OxyS RNA was detected in a Northern blot analysis during investigation of the oxidative stress regulator, OxyR (Altuvia et al. 1997). The probe used for detection of *ompR* mRNA covered the upstream region from which *oxyS* is convergently transcribed. Again, this sRNA turns out to be an antisense RNA, targeting the mRNAs of *fhlA* and, most likely, of *rpoS* as well.

The CsrA protein is a master regulator of carbon storage (and many other cellular activities). It exerts its effect by binding to target mRNAs, resulting in their degradation or translational activation. Biochemical work showed that CsrA could be found associated with an RNA (Liu et al. 1997). Direct cloning identified CsrB, an unusual RNA which carries 18 sequence (in part structural) motifs that bind one CsrA protein each. Thus, CsrB is a different kind of regulator. It does not act—as most sRNAs appear to do—by a base-pairing/antisense mechanism, but instead by protein sequestration.

2.2
Modern-Day sRNA Gene Searches

In the late 1990s, several groups embarked on genome-wide searches for sRNA-encoding genes. In 2001, three papers reported such searches, bringing the number of *E. coli* sRNAs to more than 30 (Argaman et al. 2001; Rivas et al. 2001; Wassarman et al. 2001), and subsequent searches increased this number to about 70 sRNAs that have been experimentally validated in this enterobacterium alone (see the compilation in Hershberg et al. 2003, and the references that follow). The strategies used for sRNA discovery involved different biocomputational screens, such as searches for promoter and terminator motifs within IGRs (Argaman et al. 2001; Chen et al. 2002), phylogenetic conservation analyses (Argaman et al. 2001; Wassarman et al. 2001), or RNA structure conservation between related species (Rivas et al. 2001). Some screens used detection of sRNA transcripts on high-density microarrays (Tjaden et al. 2002). One study employed shotgun cloning of sRNAs (Vogel et al. 2003), and another one identified sRNA candidates on the basis of their ability to bind the Hfq protein (Zhang et al. 2003); this protein is required for the activity of several sRNAs (Sect. 5.4). Altogether, the outcome of these screens indicated that (1) there is a significant overlap in sRNAs found in these screens, though searches are not yet saturated, (2) most sRNAs are encoded by IGRs (note, however, that most screens were biased towards these regions), (3) sRNAs are, in most cases, transcribed from their own independent genes, but some distinct sRNA species are derived from leaders or trailers of mRNAs, or otherwise processed from longer transcripts, (4) most sRNAs are indeed short (though searches allowed for longer RNAs) and often range from 60 to 150 nucleotides in length, (5) many sRNAs are transcriptionally controlled, and accumulate only under certain conditions, and (6) most sRNAs use antisense mechanisms. Finally, a survey of all sRNAs known to date indicates that sRNAs lack simple hallmarks—unlike the eukaryotic miRNAs that all share a characteristic length of approximately 22 nucleotides and are flanked by sequences that fold into recognizable precursor structures. Bacterial sR-NAs differ significantly from each other, and few, if any, sequence or structure features uniquely distinguish them from other RNAs or RNA segments.

3
The Old and New Ones—Biological Functions

The step from sRNA discovery to assignment of biological roles is a difficult one. For the 70 or so sRNAs in *E. coli*, their intracellular presence—usually under defined physiological conditions—has been confirmed by probing on Northern blots and sometimes by additional experiments, like rapid amplification of complementary DNA (cDNA) ends (RACE) and cloning as cDNA. However, functional roles could generally not be derived from the sequence or location of the gene in question. Also, the analyses of sRNA gene deletions in several laboratories gave disappointing results: growth under laboratory conditions was almost always unaffected. Thus, sRNA function must be elucidated on a case-by-case basis. For this, a number of strategies, ranging from systematic large-scale screening for downstream effects, to bioinformatics-aided predictions of targets and/or functions, can be employed (see Wagner and Vogel 2005, for a review).

In the following, we describe a few cases in which the biological roles, or at least the targets, of new sRNAs in *E. coli* have been assigned. The list is not

Table 1 List of chromosomal encoded small RNAs (*sRNAs*) in different bacterial species

Name of sRNA	Target(s)[a]	Regulatory Effect[b]	Response/ Biological role(s)
Escherichia coli sRNAs[c]			
	Housekeeping sRNAs		
4.5S RNA	SRP protein	–	Secretion
M1 RNA (RNaseP)	tRNA precursors	–	tRNA maturation
tmRNA, SsrA	–	–	Translation quality control

[a] For the antisense RNAs (all sRNAs except housekeeping RNAs, 6S RNA, RsmY and Z, CsrB and C), targets are bound in 5′ untranslated regions of the messenger RNAs (mRNAs), except when otherwise stated.
[b] (–) indicates inhibition of gene expression, (+) activation. In the case of protein targets, (–) implies protein sequestration.
[c] Many of the sRNAs in the *Escherichia coli* list have homologs in related bacteria such as *Salmonella*, *Klebsiella*, *Yersinia*, and *Erwinia* species.

Table 1 (continued)

Name of sRNA	Target(s)[a]	Regulatory effect[b]	Response/ biological role(s)
	Stress response sRNAs		
6S RNA	σ^{70} subunit of RNA polymerase	?	Stationary phase survival
CsrB–C	CsrA protein	(–)	Carbon metabolism, virulence
DicF	*ftsZ* mRNA	(–)	Cell division
DsrA	*rpoS* mRNA	(+)	Thermoregulation
	hns mRNA	(–)	
GadY	*gadX* mRNA (3′ UTR)	(+)	Acid stress
GcvB	*oppA* mRNA	(–)	Peptide transport
dppA	mRNA	(–)	
IstR	*tisAB* mRNA	(–)	SOS response
MicA/SraD	*ompA* mRNA	(–)	Membrane stress (?)
MicC	*ompC* mRNA	(–)	Membrane stress
MicF	*ompF* mRNA	(–)	Membrane stress
OxyS	*fhlA* mRNA	(–)	Oxidative stress
	rpoS mRNA	(–)	
RdlD	*ldrD* mRNA	(–)	Toxicity
RprA	*rposS* mRNA	(+)	General stress
RyaA/SgrS	*ptsG* mRNA	(+)	Glucose transport
RyhB/FerA/SraI	*sdhD* mRNA	(–)	Iron homeostasis
	sodB mRNA	(–)	
Spot42	*galK* mRNA	(–)	Galactose metabolism
Sok-like RNAs	*hok-like* mRNAs (?)	(–)	Nutritional stress (?)
Streptococcus pyogenes			
FasX and Pel	cell-suface proteins (?)	(+)?	Virulence
Staphylococcus aureus			
RNA III	*spa* mRNA	(–)	Virulence
	hla mRNA	(+)	
Pseudomonas aeroginosa			
PrrF1–2	*sodB* mRNA (?)	(–)	Iron homeostasis
Pseudomonas fluorescens			
RsmY–Z	RsmA, RsmE proteins	(–)	Production of secondary metabolites
Vibrio cholerae			
Qrr1–4	*luxR* mRNA	(–)	Quorum sensing,
	hapR mRNA	(–)	virulence

[a] For the antisense RNAs (all sRNAs except housekeeping RNAs, 6S RNA, RsmY and Z, CsrB and C), targets are bound in 5′ untranslated regions of the messenger RNAs (mRNAs), except when otherwise stated.

[b] (–) indicates inhibition of gene expression, (+) activation. In the case of protein targets, (–) implies protein sequestration.

complete, and additional information can be found in reviews (Altuvia 2004; Altuvia and Wagner 2000; Gottesman 2004; Gottesman et al. 2001; Storz et al. 2004; Vogel and Wagner 2005; Wagner and Flärdh 2002; Wagner and Vogel 2003, 2005; Wassarman 2002). As expected, the majority of these belong to the "class" of antisense RNAs, and involvement in stress responses emerges as a common theme (Table 1).

3.1
Many sRNAs Are Involved in Stress Response Regulation

Expression patterns showed early on that only a few of the new sRNAs are present under all (or at least the ones tested) growth conditions. Instead, many are induced when cells enter the stationary phase but are not transcribed in the logarithmic growth phase. Entry into the stationary phase is correlated with a complex series of adaptive changes in which several stress responses are triggered. The specific induction of sRNA synthesis in this growth phase suggests that sRNAs might be acting under these conditions, possibly as regulators. Other sRNAs accumulate under cold shock treatment, iron depletion, onset of the SOS response, or sugar stress. Thus, a priori, it can be expected that many, or even most, sRNAs are involved in stress response regulation.

Some of these sRNAs provide good examples for working protocols to assign functions. In some cases, binding sites (motifs) for known regulatory proteins, near sRNA gene promoters, permitted "educated guesses" as to what regulatory pathway an sRNA may be part of.

3.1.1
MicF, MicC, and MicA

E. coli encodes a number of abundant outer-membrane proteins. Several of these are trimeric porins that control the influx of low molecular weight compounds; some may have additional roles in virulence, membrane integrity, etc. Two porins, OmpF and OmpC, are regulated in various stress situations, generally in opposite directions, presumably to adapt membrane properties to the conditions at hand; OmpF pores and OmpC pores differ in diameter. Both genes, *ompF* and *ompC*, are under complex transcriptional regulation. In addition, their translation is regulated by two antisense RNAs, MicF and MicC, respectively. MicF—the first chromosomally encoded antisense RNA found in *E. coli*—is co-regulated with *ompC*. Thus, when *ompC* transcription is on, MicF accumulates as well (Andersen et al. 1987). This sRNA then base-pairs to its target, the *ompF* mRNA, to block its translation (Andersen and Delihas 1990). Thus, OmpC levels rise, and OmpF levels are decreased. More recently, the Storz group showed that a new sRNA, MicC, is responsible for post-transcriptional inhibition of *ompC* mRNA, essentially by the same mechanism as shown for MicF/*ompF* (Chen et al. 2004). This work also showed that expression of MicF and MicC is almost always mutually exclusive; when the MicC concentration is high, that of MicF is low—in line with the requirement for opposite regulation of the two porins.

Another abundant outer-membrane protein, OmpA, becomes downregulated upon entry into the stationary phase. This regulation had earlier been attributed to direct effects of the binding of Hfq to the leader of the *ompA* mRNA (Vytvytska et al. 2000). Recent work has shown that this Hfq effect is indirect. A new sRNA, MicA (previous name SraD; Argaman et al. 2001), accumulates to high levels in the stationary phase, where it acts as an anti-

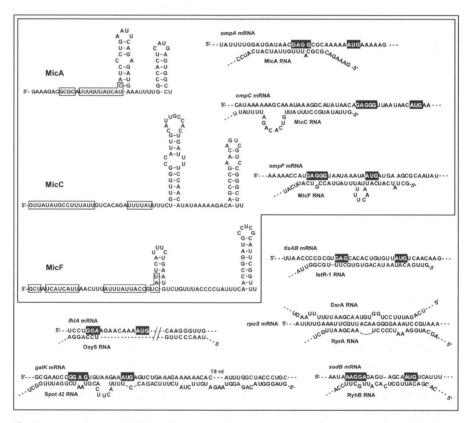

Fig. 1 Some antisense RNAs and antisense/target RNA complexes. *Upper box, left*: MicA, MicC, and MicF RNA secondary structures. Nucleotides involved in target interaction are *boxed*. *Upper box, right* and *lower box*: Examples of base-paired regions known (or predicted) to be formed between several antisense RNAs and their respective target RNAs. *Black boxes with white letters* indicate the Shine–Dalgarno (*SD*) and AUG start codon sequences in the messenger RNAs (*mRNAs*). In *rpoS* mRNA, the SD/AUG region lies downstream (see text for details), and interactions with RprA as well as DsrA are indicated (below and above the target sequence)

sense inhibitor of *ompA* mRNA. In vivo, this regulation becomes impaired when Hfq is absent; deletion of the *micA* gene leads to loss of *ompA* downregulation (Udekwu et al. 2005). Like *ompF* and *ompC*, the *ompA* gene is under both transcriptional and post-transcriptional (antisense RNA) control. Thus, of the outer-membrane protein family members, three—so far—have cognate antisense RNAs that specifically inhibit their respective target genes, each under appropriate stress conditions. In all three cases, Hfq is required for antisense regulation, but a mechanistic understanding of its action is lacking. Base-pairing interactions of all three antisense-target RNA pairs are shown in Fig. 1 and illustrate the recurrent theme that TIRs are preferred target sites. A further interesting question arising from a comparison of the three Mic RNAs concerns specificity and will be dealt with later.

3.1.2
OxyS

OxyS has been described as a regulator of as many as 40 genes, most of them part of the oxidative stress regulon. This sRNA is under the control of the OxyR protein and is induced by, for example, peroxide treatment (Altuvia et al. 1997). Two targets are known: *fhlA* (encoding a transcriptional activator) and *rpoS* (Altuvia et al. 1998; Zhang et al. 1998). In the case of *fhlA* mRNA, it has been shown that OxyS base-pairs to two noncontiguous regions within the TIR (Fig. 1), preventing translation. Since this target RNA encodes a regulatory protein, the OxyS effects on several downstream genes may be indirect. OxyS-dependent inhibition of *rpoS* expression was initially attributed to titration of Hfq (this protein is required for efficient translation of RpoS), but also appears to involve a conventional antisense mechanism.

3.1.3
DsrA and RprA

The *rpoS* gene encodes σ^S, a key player in adaptive responses to various stresses (Hengge-Aronis 2002). Regulation of *rpoS* is complex and occurs by several mechanisms. In particular, several pathways of stress regulation converge on *rpoS*, and at least three sRNAs have roles in this (Repoila et al. 2003). One of these is OxyS (Sect. 3.1.2). DsrA and RprA are two additional sRNAs that act on the same target, and do so by base-pairing to overlapping but nonidentical sites within the *rpoS* mRNA (Lease et al. 1998; Majdalani et al. 2002; Sledjeski et al. 1996) (Fig. 1). Interestingly, RpoS translation is stimulated rather than inhibited by interaction with these sRNAs, and Hfq is required. Apparently, the *rpoS* mRNA folds into a translation-incompetent secondary structure in which the TIR is inaccessible owing to intramolecular base-pairing. DsrA and RprA can base-pair to the self-inhibitory upstream half of this sequence to free the ribosome loading site, thus activating translation. What is the rationale for using two (or more) sRNAs on the same target? The answer probably lies in the expression patterns of the sRNAs. DsrA accumulates at low temperature (Repoila and Gottesman 2003), RprA is controlled by the RcsC/RcsB phosphorelay system which also controls capsule synthesis (Majdalani et al. 2002), and OxyS is induced by oxidative stress (Altuvia et al. 1997); hence, it is likely that sRNAs serve to feed information on specific stresses into regulation of global stress responses.

As mentioned in Sect. 2.1, DsrA has an additional function. It acts as antisense RNA on *hns* (global transcriptional regulator) mRNA and thus affects expression of the genes that are under the control of H-NS. The two target-complementary segments of DsrA are distinct, so that one sequence exclusively targets *rpoS* (to activate) and the other *hns* (to inhibit).

3.1.4
Spot 42

Spot 42 was discovered more than 30 years ago, but has only recently been functionally assigned. The Valentin-Hansen group demonstrated that Spot 42 is under cyclic AMP (cAMP) receptor protein CRP–cAMP control. When the glucose concentration is high, induction of this sRNA results in decreased levels of the GalK (galactokinase) protein, encoded by the third cistron of the *galETKM* operon (Møller et al. 2002b). Base-pairing of Spot 42 to the TIR of *galK*, aided by Hfq, results in inhibition of translation (Møller et al. 2002a). This leads to discoordinate expression from the *gal* operon, since antisense inhibition does not affect the two upstream reading frames. The rationale for this regulation is not entirely clear, but is likely connected to control of carbon utilization.

3.1.5
RyhB

RyhB/SraI was identified in two screens (Argaman et al. 2001; Wassarman et al. 2001). Inspection of its promoter region identified a FUR site, known to bind the Fur repressor which is active at high intracellular Fe^{2+} concentrations. Work by the Gottesman group solved a puzzle: Fur was known to *down*regulate genes involved in iron assimilation, but also to *up*regulate certain iron-containing and iron-storage proteins—although it is generally regarded as a repressor. It turns out that RyhB becomes induced (derepressed) when Fur becomes inactive (low iron). The sRNA then accumulates to high levels and, by an antisense mechanism that also requires the intracellular presence of the Hfq protein, targets at least two mRNAs of iron-containing proteins, *sodB*(superoxide dismutase) and *sdhD*(succinate dehydrogenase subunit) (Massé and Gottesman 2002; Vecerek et al. 2003). Similar to Spot 42, the sRNA promotes discoordinate expression of genes within an operon; the target sequence of RyhB is located across the TIR of the second cistron of the *sdhCDAB* operon (Massé and Gottesman 2002). Thus, under high iron conditions, Fur indirectly activates the *sodB* and *sdhD* genes by repressing the transcription of RyhB. Several additional target mRNAs for RyhB have been suggested but await further experimental validation.

3.1.6
IstR

IstR is involved in yet another stress: DNA damage/the SOS response. The *istR* locus is located immediately upstream of the *tisAB* operon (previously denoted *ysdAB*, and at that time of unknown function). A bidirectional LexA (repressor of SOS response genes) binding site is located between the divergently oriented *istR* and *tisAB* genes. It turns out that *tisAB* (toxicity induced by SOS) encodes an SOS-induced toxin, TisB, which arrests bacterial growth (Vogel et al. 2004). The *tisA* reading frame, which—unlike that of *tisB*—is not conserved, is not translated (Unoson et al. unpublished results). The *istR* (inhibition of SOS-induced toxicity by RNA) locus encodes two sRNAs; they are transcribed from two independent promoters and share a terminator. Synthesis of the shorter RNA, IstR-1, is constitutive. IstR-1 carries 21 nucleotides that are complementary to a segment of *tisAB* mRNA. Interestingly, the full-length *tisAB* mRNA is translationally inert, and requires activation by endonucleolytic cleavage 40 nucleotides from the 5′-end. When IstR-1 is present in high excess over its target, it binds efficiently to its target site, entailing RNAse III cleavage. This in turn converts the *tisAB* mRNA to an RNA species which is inactive, both in vitro and in vivo, although it contains the entire *tisB* reading frame. Thus, the function of IstR-1 is to prevent inadvertent toxicity when the SOS response is off. In contrast, when the SOS response is triggered, accumulation of *tisAB* mRNA overrides this protection. IstR-1 is titrated out, and cleaved by RNAse III in an interaction-dependent fashion. Cell growth slows down and physiological adaptation and DNA repair ensue. The second sRNA, IstR-2, is under LexA control and is thus induced in parallel with *tisAB*. Interestingly, although IstR-2 shares the entire region of complementarity to *tisAB* with IstR-1, it is inefficient in binding and protecting from toxicity. Thus, it is expected that, under SOS conditions, IstR-2 targets another yet unidentified mRNA.

3.1.7
CsrB and CsrC

CsrB and CsrC (and their relatives in other bacteria) are regulators of a different kind. They act by protein sequestration. Their target is the very abundant CsrA protein, a regulator of glycogen synthesis and catabolism, as well as of cell-surface properties, motility, and biofilm formation (Romeo 1998). When

CsrB is induced and accumulates, each molecule binds up to 18 CsrA proteins, thus indirectly upregulating genes which otherwise are translationally repressed by CsrA binding (e.g., *glgC*, encoding a glucose 1-phosphate adenylyltransferase subunit). CsrB carries multiple exposed sequence motifs (variants of CAGGAUG) which compete for CsrA binding to similar sequences located in the TIRs of target RNAs (Liu et al. 1997). Additional complexity is evident in an autoregulatory loop; *csrB* gene transcription is upregulated by CsrA, and is also under the control of the BarA (histidine kinase)/UvrY (response regulator) two-component system (Suzuki et al. 2002). The role of CsrC is less clear, though its action on CsrA indicates that—similar to what has been found in other bacteria—it supplies redundancy (Weilbacher et al. 2003). The fact that *csrA* homologs can be found in approximately 70 diverse bacterial species may indicate that CsrB-like RNAs will also be identified—as they already have been in *Pseudomonas, Erwinia, Salmonella, Shigella, Legionella, Serratia, Helicobacter,*and some other bacteria. RNAs able to titrate a regulatory protein may be a recurrent theme at least for homeostatic control of carbon fluxes and related activities.

3.1.8
Others

DicF is a small, processed sRNA which is encoded by a defective prophage (Tétart and Bouché 1992). A partial region of complementarity promotes interaction with the *ftsZ* mRNA (encoding the major cell division protein). This results in inhibition of cell division, though the biological implications of this are not understood. **GcvB** was found serendipitously during studies of the *gcv* system (glycine cleavage). This sRNA downregulates two target mRNAs, *opp* and *dpp* (Urbanowski et al. 2000). These genes encode components of two major peptide transport systems. Mechanistic understanding is still lacking. **SgrS** (previously RyaA), a recently discovered Hfq-binder sRNA, is involved in the phosphosugar stress response (Vanderpool and Gottesman 2004). This RNA is transcriptionally activated by the SgrR (previously YabN) protein. When glucose 6-phosphate is at high concentration, the *ptsG* gene (encoding EIICB^{Glc}, the membrane component of the glucose transporter) is post-transcriptionally downregulated. Control requires Hfq, RNase E, and SgrS, and ultimately results in target RNA degradation. Base-pairing between the sRNA and its target has been suggested but awaits experimental support. The Aiba group recently showed that SgrS access may be aided by membrane localization of the target mRNA (Kawamoto et al. 2005). **GadY**, identified in a previous screen (Chen et al. 2002), is transcribed from the IGR between *gadX* and *gadW* (involved in the acid response). The orientation of the *gadY* gene is opposite to that of *gadX/gadW*. Thus, in the stationary phase when GadY is induced in a σ^S-dependent fashion, this sRNA acts on the 3' untranslated region (UTR) region of the *gadX* mRNA (Opdyke et al. 2004).

Interestingly, this results in stabilization of the target mRNA and, hence, upregulation. GadY is one of the few cases of *cis*-encoded chromosomal sRNAs (Sect. 4.1). Other known *cis*-encoded sRNAs are the **Sok** homologs. In plasmids like R1, *hok/sok* loci confer stability of maintenance by postsegregational killing of plasmid-free cells (Gerdes et al. 1997). Hok is a toxin that kills cells. Sok is the antitoxin (antisense RNA) that prevents killing by blocking Hok translation. Owing to its short half-life, Sok is rapidly degraded when the plasmid is lost, and cells die. The Gerdes group has reported the presence of many *hok/sok*-like loci in bacterial chromosomes (Pedersen and Gerdes 1999). At this point, their roles are unknown. Several of these *hok/sok* loci have suffered inactivation by insertion sequences. Nevertheless, at least three chromosomal loci in different *E. coli* strains express Sok RNAs. Another toxin–antitoxin system is present in four copies, encoded by the so-called long direct repeat (LDR) sequences. One of these, LDR-D, encodes a toxin, LdrD (Kawano et al. 2002). Killing is counteracted by a *cis*-encoded antisense RNA, **RdlR**. Why these (and other) killer systems exist in bacteria is still enigmatic, though an involvement in nutritional stress responses is likely (Christensen et al. 2003).

4
What's Out There in Other Bacteria?

Clearly, sRNAs are not only encoded by *E. coli*. The high number of known sRNAs in this bacterium and the comparably small harvest in others simply reflect the many systematic searches conducted in *E. coli*. First, it is clear from the conservation criterion used in bioinformatics screens that near relatives of *E. coli* harbor the homologous genes for the majority of the sRNAs. Certainly, however, some sRNAs have specifically evolved in other bacterial lineages, and some will even be species-specific.

4.1
Regulation of Virulence

In some Gram-positive bacteria, sRNAs play major roles in the regulation of pathogenesis-/virulence-related genes.

4.1.1
RNAIII, a Key Regulator of Virulence in *Staphylococcus Aureus*

RNAIII was discovered as a global virulence regulator in *Staphylococcus aureus* more than 10 years ago (Morfeldt et al. 1995; Novick et al. 1993). This RNA, which is part of the *agr* (accessory gene regulator) locus, has some remarkable properties. Its gene is under the control of a quorum sensing system. When the cell density increases, RNAIII accumulates (peaking in the late exponential phase) and affects many downstream genes. In particular, it upregulates genes involved in secreted toxin and enzyme production, whereas it downregulates genes that encode cell-surface proteins. As of today, we know of three different functional modules within this relatively large approximately 514-nucleotide RNA. One module is a short reading frame encoding *hld* (hemolysin δ); i.e., RNAIII functions as an mRNA. A second, structural domain near the 5′ end acts as an antisense RNA. It base-pairs to a region within the *hla* (hemolysin α) mRNA, effectively in competition with a leader segment that otherwise sequesters the *hla* TIR (Morfeldt et al. 1995). Thus, RNAIII activates *hla* translation. The third target gene is *spa* (protein A, a cell-surface protein). Recent work has shown that a 3′-segment of RNAIII base-pairs to the TIR of the *spa* mRNA, inhibiting protein A translation (Huntzinger et al. 2005). Thus, RNAIII is an mRNA, an *activator* antisense RNA, and an *inhibitory* antisense RNA—all activities contained within one RNA molecule. There may be more to come; previous data suggested that RNAIII may also have effects on the transcription of certain target genes, though, at this point, experimental support is lacking.

4.1.2
Virulence-Associated RNAs in *Streptococcus Pyogenes*

Regulation of virulence in group A streptococci (GAS) is equally complex, involving several signal transduction systems. Similar to *S. aureus*, regulatory RNAs may play key roles, though the mechanisms of regulation are in this class of bacteria not yet understood in detail. Two sRNAs, FasX (Kreikemeyer et al. 2001) and Pel (pleiotropic effect locus) RNA (Mangold et al. 2004), are encoded at loci known to be required for virulence. Reminiscent of RNAIII (Sect. 4.1.1), these sRNAs are unusually long for typical riboregulators and contain short ORFs. For example, Pel RNA encodes *sagA* (streptolysin S), but mutational analyses demonstrated that the regulatory function resides in the RNA itself. The induction of the sRNA is stimulated by conditioned medium, and thus is likely under quorum sensing control. When induced, Pel RNA activates the synthesis of surface-associated and secreted proteins (Mangold et al. 2004), but it is yet unclear whether this occurs by a transcriptional or a post-transcriptional mechanism (i.e., antisense).

4.2
sRNAs and Quorum Sensing

Studies of *Vibrio harveyi* and *V. cholerae* uncovered several sRNAs that regulate genes involved in intercellular (cell-density-dependent) communication; quorum sensing controls bioluminescence in *V. harveyi* and virulence and biofilm formation in *V. cholerae*. The Bassler group conducted a mutant screen for target gene (*lux* genes) regulators that act downstream of the LuxO-P protein (response regulator, which requires σ^{54}), on which cell density sensing signals converge in both bacteria (Lenz et al. 2004). This screen identified *hfq*. Since Hfq in other bacteria is often required for antisense-target RNA control circuits, this suggested the possibility that the defect in the *hfq* mutant strain reflects the involvement of sRNAs in this phenotype. Indeed, a bioinformatics search identified four similar sRNAs, Qrr1–Qrr4 (quorum regulatory RNA), which were shown to be induced by LuxO-P/σ^{54}. Although interactions with their putative target RNAs have not been experimentally tested, it appears that these sRNAs, aided by Hfq, act as antisense RNAs to post-transcriptionally downregulate the *luxR* (*V. harveyi*) and *hapR* (*V. cholerae*) master regulator genes.

4.3
Other Non *E. coli* sRNAs

Some sRNAs that are known to be present in *E. coli* are also encoded by other bacteria. Many of these have been tentatively identified as homologs, others have been functionally characterized. A good example of this is provided by the functional equivalents of the *E. coli* CsrB and CsrC RNAs (Sect. 3.1.6). The primary sequences of these genes in different bacteria are so divergent that it is unclear whether they are bona fide homologs, though their regulatory function, via interaction with CsrA-like regulators, is conserved. For example, the RsmZ and RsmY RNAs of *Pseudomonas fluorescens* CHA0—which are much shorter and quite dissimilar to their *E. coli* counterparts—carry several AGG sequence motifs that sequester a CsrA-homologous regulator protein, RsmA (regulator of secondary metabolism) (Valverde et al. 2004). As in *E. coli*, the regulatory circuit also involves a two-component system, here GacA–GacS. The PrrB RNA probably carries out a similar role in *P. fluorescens* F113 (Aarons et al. 2000). Finally, *P. aeruginosa* encodes two RNAs, PrrF-1 and Prrf-2, that lack homology to *E. coli* RyhB, but appear to be functionally equivalent, regulating iron homeostasis (Wilderman et al. 2004).

5
Mechanisms: Antisense and Protein Sequestration

From a survey of the (old and new) sRNAs that have been characterized in *E. coli*, it is clear that (1) most of them are gene regulators and (2) they do so by antisense mechanisms. The exceptions to this—apart from the "housekeeping" RNAs M1 of RNase P, 4.5S RNA, and tmRNA—are CsrB/CsrC and possibly 6S RNA. CsrB/CsrC and their regulatory role by protein sequestration were described in Sect. 3.1.7. 6S RNA is an abundant, highly structured RNA which shows affinity to the σ^{70} subunit of RNA polymerase (Wassarman and Storz 2000). Since no clear phenotypes have been associated with over-expression or gene deletion, the functional significance of this RNA is still elusive (see, however, Trotochaud and Wassarman 2004).

Much more is known about antisense RNA mechanisms in bacteria. Most of the earlier work has provided in-depth knowledge on several plasmid control systems (for a review, see Wagner et al. 2002), and substantial information is already available on some chromosomally encoded systems. Aspects of the latter are discussed later.

5.1
Antisense: *cis*-Versus *trans*-Encoded

Most sRNAs act through antisense mechanisms; they regulate by base-pairing to target RNAs (Fig. 2). With a few exceptions, the chromosomally encoded sRNAs differ from the antisense RNAs of plasmids, transposons, and phages in one important property. In the last cases, the antisense RNAs are *cis*-encoded: transcription of antisense and target RNA occurs from the same DNA segment but with opposite polarity. This implies that antisense and target RNA exhibit complete complementarity, and that targets are unique. Most chromosomal sRNA genes are *trans*-encoded. They can (but do not have to) be located in the vicinity of their target genes, but are not encoded by the same DNA segment. As a consequence, there is no full complementar-

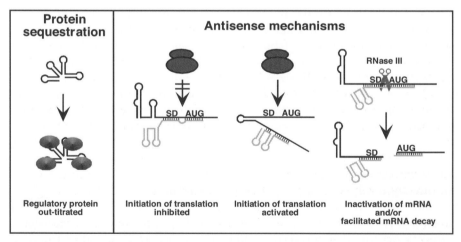

Fig. 2 Some models of small RNA (**sRNA**) dependent regulation. *Left* to *right*:Regulatory RNA sequesters a protein (e.g., CsrA); binding of an antisense RNA inhibits translation by occluding ribosome access (e.g., MicA); antisense RNA binding activates translation (e.g., DsrA); antisense/target RNA pairing, if long enough, creates a substrate for RNase III, resulting in functional inactivation or facilitated decay (e.g., IstR-1). Antisense RNAs are in *gray*. Consult the text for examples of these mechanisms

ity; base-pairing of antisense and target sequences is limited, noncontiguous, and noncanonical intermolecular base-pairs are often encountered. It may be precisely because of these features that chromosomally encoded sRNAs have been able to evolve into regulators of multiple target RNAs (Sects. 3.1.2, 3.1.3).

Even though we know of too few cases to reach a clearcut conclusion, an additional distinction between *cis*- and *trans*-encoded antisense RNAs is suggested. Many bacterial sRNAs require the Sm-like Hfq protein as a helper (Sect. 5.4), whereas plasmid-encoded (*cis*-encoded; fully complementary to target) antisense RNAs do not (Slagter-Jäger and Wagner, unpublished results). If this pattern were to be supported in future studies, it might imply that the evolutionary path to regulatory properties is different. Mutational "testing" in *cis*-encoded antisense/target systems rarely leads to loss of control (since any base change in the DNA maintains complementarity between the two interactants), and thus maximized RNA–RNA binding rates as well as independence from helper proteins could be selected for. In *trans*-encoded antisense/target systems, sequences can drift substantially over evolutionary time, as long as regulatory function is not compromised. In accord with this, only the sites of complementarity are most often strictly conserved, as is the location of the recognition site in the target, whereas neighboring sequences tend to show divergence when related species are compared.

5.2
Antisense: Sites of Action—Common Themes

A survey of known interaction sites strongly suggests that ORF-internal sequences are rarely targets. This makes sense, since tightly spaced moving ribosomes should not allow for easy sRNA access. Even if binding were to occur, ribosomes are able to unwind secondary structure elements and thus are expected to strip sRNAs off their target sequence. Indeed, most (antisense) sRNAs with known or predicted targets appear to base-pair to 5′-UTRs in the vicinity of, or overlapping, TIRs. Of the genetically (by mutations) or biochemically (by in vitro interaction/probing analyses) validated cases, this is true for MicF, MicC, MicA, RyhB, DsrA, RprA, IstR, OxyS (Fig. 1 and references in Sects. 3.1.1–3.1.4, 3.1.6), and for most predicted ones. This makes perfect sense if the "purpose" is to affect translation. Base-pairing often directly involves nucleotides that otherwise base-pair to ribosomal RNA and initiator tRNA, respectively—the Shine–Dalgarno (SD) sequence and the AUG (or GUG, UUG) start codon. In other cases the RNA duplex forms immediately upstream, without overlap of SD/AUG. However, secondary structures (hairpins/intramolecular duplexes) immediately upstream of a SD are known to completely abolish translational initiation, in vitro and in vivo (Malmgren et al. 1996), and thus sRNA binding at either site has the same full inhibitory potential. GadY instead targets a 3′-UTR of the *gadX* mRNA, but here the result is stabilization of the mRNA (Sect. 3.1.8).

If TIRs are preferred targets for sRNAs, then suitable interaction sites should be present in a defined structural context—as is true for the many plasmid antisense RNA cases (Wagner et al. 2002). In line with this, all the target RNAs for the previous list of sRNAs carry relatively long 5′-UTRs which, wherever experimentally tested, indeed fold into distinct secondary structures. Accessibility of interacting sequences is another important requirement for control. The complementary sites in the interacting RNAs should be at least partially single-stranded, so that base-pairing can start. Certainly, protein factors may facilitate this step by modulating/unfolding RNA structures (Sect. 5.4).

5.3
Antisense: Mechanisms of Inhibition—Common Themes?

Mechanistic explanations for sRNA-dependent regulation have been proposed; in many cases, it is assumed that Hfq-aided sRNA binding (Sect. 5.4) causes mRNAs to be degraded in an RNase E dependent fashion. Several papers convincingly show that indeed target mRNA half-lives are decreased upon induction of the cognate sRNAs. Whether this necessarily implies that promotion of RNase E cleavage reflects functional inactivation, or rather is its consequence, remains unclear (Carpousis 2003). Massé et al. (2003) showed that the interaction of RyhB and the *sodB* mRNA in the cell results in "coupled degradation" of both RNAs. The authors suggested that the formation of the sRNA–mRNA complex could create substrates for RNase E.

Distinguishing cause and effect is not easy when it comes to degradation models. For instance, when an sRNA base-pairs to a TIR, ribosome binding is prevented. Thus, unless RNA–RNA interaction is highly reversible, the mRNA is at this point functionally inactivated, whether or not it persists as an abundant species in the cell. Transcription elongation of a translationally blocked nascent mRNA (transcription and translation are coupled in bacteria) will generate naked mRNA which most often decays at an accelerated rate. This is due to the increased accessibility of internal sites to ribonucleases, and may or may not involve RNase E as the rate-limiting nuclease. Concerning the observed effects of RNase E, RNase E cleavage sites have been mapped in the vicinity of TIRs in several mRNA leaders (Vytvytska et al. 2000). Thus, initiating ribosomes may protect an mRNA from RNase E promoted degradation by competition for this site(s). In this scenario, an sRNA which in turn competes with initiating ribosomes may "deprotect" the RNase E site, and decay—involving RNase E—will ensue. In vitro studies in our group have shown that MicA binding to the *ompA* mRNA leader has no effect on RNase E cleavage. Instead, the sRNA prevents ribosome binding (Udekwu et al. 2005). Thus, in our opinion, the role of RNase E in *functional* inactivation of mRNAs, when bound by sRNAs, is still open and requires detailed in vitro analyses.

5.4
Antisense: Protein Helpers Required, And for What?

Most if not all *cis*-encoded antisense RNAs from plasmids manage to bind to, and inhibit, their target RNAs without a requirement for helper proteins. By contrast, sRNAs in *E. coli* can often be isolated from cells in complexes with the ubiquitous RNA binding protein Hfq (Wassarman et al. 2001; Zhang et al. 2003), suggesting that this protein is involved in their functions (Valentin-Hansen et al. 2004). In line with this expectation, mutations in the *hfq* gene most often lead to loss of sRNA function. So what does Hfq do, and why is is required?

Hfq was initially identified as a host factor needed for $Q\beta$ (RNA phage) replication. Genes encoding *hfq* are present in many diverse bacteria, but are lacking from others. The *hfq* gene is dispensable, but mutants display pleiotropic phenotypes and even loss of virulence in pathogenic bacteria. The Hfq protein is abundant in exponentially growing *E. coli* at approximately 60 000 copies per cell and decreases threefold in the stationary phase (Ali Azam et al. 1999). Electron microscopy and crystal structure determination have shown that it forms doughnut-shaped homohexameric rings. The inner face of the torus binds RNA as seen in co-crystals with a short RNA model substrate (Schumacher et al. 2002). Hfq belongs to the family of eukaryotic Sm and Sm-like proteins that interact particularly with U RNAs and are important for splicing; Sm proteins form heteroheptamers. Hfq is an avid RNA binder, can affect polyadenylation, translational efficiency, and RNA stability, and has been proposed to act as an "RNA chaperone" (Moll et al. 2003). Its important role in many sRNA-mediated regulatory interactions is undisputed, but mechanisms have remained enigmatic.

Like the Sm proteins, Hfq binds preferentially to A/U-rich unstructured RNA regions, often near stem loops (Brescia et al. 2003). Such motifs are often encountered in sRNAs (and in several target RNAs), and Hfq binds to many RNAs with Kd values in the nanomolar or even subnanomolar range. How can strong binding enhance effects of sRNAs on target RNA regulation? One possibility is stabilization of sRNAs. Hfq binding can protect some sRNAs from endonucleolytically initiated decay (e.g., by RNase E). Thus, the intracellular sRNA concentration would be higher in an hfq^+ than an hfq^- strain (e.g., SgrS, Kawamoto et al. 2005; Spot 42, Møller et al. 2002a; DsrA, Sledjeski et al. 2001; GadY, Opdyke et al. 2004; MicF, Slagter-Jäger and Wagner, unpublished results); in principle, higher concentrations of a regulator might account for the differences in regulation observed. Some other sRNAs, for example, OxyS (Zhang et al. 2002) and IstR-1 (Unoson and Wagner, unpublished results) are not stabilized by Hfq. Yet, at least OxyS depends on this protein for *activity*. Others may be affected in stability *and* activity. On the basis of in vitro studies of the OxyS/*fhlA* and Spot 42/*galK* systems it was proposed that Hfq can "enhance" sRNA/target RNA binding (Møller et al. 2002a; Zhang et al. 2002).

Whether this reflects an on rate or a Kd effect is at present unknown. Another plausible mechanism by which Hfq could aid sRNAs involves structural changes in one or both of the interactant RNAs. For example, unfolding of a stem region would facilitate intermolecular pairing since a complementary site would become available (Geissmann and Touati 2004; Zhang et al. 2002), which would be equivalent to an increase in the concentration of binding-competent RNAs. The structure of DsrA seems unaffected by Hfq (Brescia et al. 2003), whereas that of, for example, RyhB is affected (Geissmann and Touati 2004). A variant of these models postulates that Hfq-bound sRNAs and target RNAs could be brought together by interaction between Hfq hexamers, i.e., a molecular crowding effect. At this point, the mechanism(s) by which Hfq facilitates sRNA-dependent regulation is unclear, and a careful study of DsrA/$rpoS$ mRNA binding revealed a mere twofold stimulation of the association rate by Hfq (Lease and Woodson 2004).

It is worth noting that several other abundant global regulators share RNA-binding properties. StpA, HU, and H-NS all bind RNAs. For instance, StpA, an RNA chaperone (Waldsich et al. 2002) affects MicF stability in vivo (Deighan et al. 2000). However, no studies have yet investigated the possible involvement of these proteins in sRNA-mediated regulation.

5.5
Antisense: the Specificity Problem

All antisense RNAs must base-pair to target RNAs to be effective regulators. The simplicity of this statement hides several problems: How many base-pairs are needed for stable binding? Is regulation a function of the stability of a complex (Kd) or the association rate? How important is the effect of sRNA and target RNA structure on binding and regulatory efficiency? Studies of plasmid copy number control systems, like ColE1 and R1, have shown that the association rate is the most important determinant of an antisense RNA. Yet, it is often assumed that the stability of an RNA duplex is the critical parameter.

The stability of an RNA duplex is described by its free energy (ΔG°), which in turn is related to the equilibrium dissociation constant Kd. A Kd tells us the fraction of the two interactant RNAs in the free versus the bound state, and is related to the half-life of the complex under a given set of conditions (temperature, salt concentration). If one tests sets of pairwise random genomic sequences (as RNA) of length 50 and 40, respectively, predicted base-pairing (Hodas and Aalberts 2004) is often noncontiguous and comprises a total of about 13–19 base pairs, giving ΔG° values ranging from -8 to $-13\,\mathrm{kcal\,mol^{-1}}$ (data not shown). A look at (in part) experiment-based antisense/target RNA complexes (Fig. 1) indicates much lower ΔG° values. The Mic RNAs, for example, have values of approximately $-25\,\mathrm{kcal/mol^{-1}}$, RyhB/$sodB$ is at -17, DsrA/$rpoS$ at -28, RprA/$rpoS$ at -24, and Spot 42/$galK$ may be as low as $-48\,\mathrm{kcal/mol^{-1}}$. It makes intuitive sense that "real" binders display higher stability of the RNA–RNA complex than do random RNA pairs, and it appears that selection has driven such interactions far beyond background values.

Conversely, does a predicted ΔG^o value of, say, – 20 kcal/mol^{-1} imply that the RNAs in question will interact? Not necessarily, since this requires that the binding *pathway* permits the RNAs to arrive at this "final" state. The plasmid antisense/target RNA cases (e.g., CopA/CopT of plasmid R1: Kolb et al. 2000; Nordgren et al. 2001; Nordström and Wagner 1994) show that single base mutations that have no effects on a predicted Kd, but that affect the association rate, can abolish regulation entirely. For the *trans*-encoded systems, the IstR-1/IstR-2 antisense RNAs can serve as an example (Vogel et al. 2004). All target-complementary nucleotides of IstR-1 are present in IstR-2, but the latter binds the *tisAB* mRNA inefficiently, and is a poor inhibitor of TisB translation. This cannot be reconciled with a Kd effect. Instead, *structural accessibility* must play a role (Hjalt and Wagner 1995; Slagter-Jäger and Wagner 2003). The longer IstR-2 carries approximately 65 extra nucleotides, a few of which sequester several nucleotides that are part of the single-stranded 5′-tail in IstR-1 (Fig. 3). As a consequence, the binding rate drops significantly. Since a complex between two RNAs must be able to initiate without structurally impeding this step (which involves nucleation and rapid formation of a number of neighboring base-pairs), sufficiently unstructured regions must be present in both interacting RNAs at sites of sequence complementarity. Structural sequestration of sequences needed for interaction can prevent binding or slow it down, despite potential base-pairing capacity.

A related problem concerns how specificity is generated. If one conducts searches for reasonable target sites of some sRNAs (by base-pairing/ΔG^o criteria), putative cross-reactions often show up. For example, Fig. 4 shows predicted base-pairing for noncognate Mic/*omp* mRNA combinations, in comparison with their cognate interactions, and with those of a heterologous pair set (Spot 42 as the sRNA). The ΔG^o values for the Spot 42/*omp* RNA interactions range from – 10 to – 13 kcal/mol^{-1} (weak binding), whereas the heterologous Mic/*omp* RNA pairs exhibit respectable predicted stabilities; two thirds would have *dissociation rate constants* from 10^{-8} to 10^{-9} s^{-1} (given that the association rate constants are generally approximately 10^6 M^{-1} s^{-1} (Wagner et al. 2002)). Thus, if formed, these RNA–RNA complexes would remain undissociated for more than an average cell generation, i.e., these Mic RNAs should therefore *also* regulate heterologous *omp* mRNAs. This is not the case. Each Mic RNAs is specific, and overexpression virtually abolishes synthesis of the protein encoded by the cognate mRNA without affecting the output of the other two proteins. The same specificity is supported by in vitro binding experiments (Darfeuille and Wagner, unpublished results).

In summary, ΔG^o values are rarely good predictors for (unknown) RNA–RNA interactions since binding *rates* are strongly affected by *how* one arrives at the complex—and this must take accessibility and structural presentation

Cognate combinations

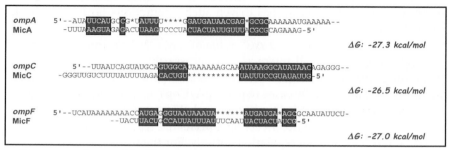

Near-cognate combinations

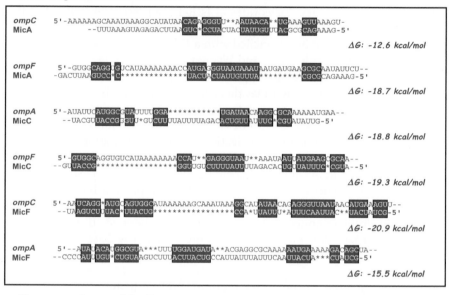

Non-cognate combinations

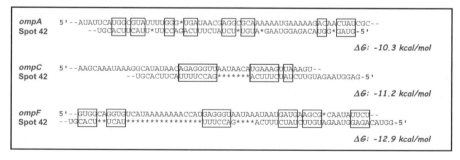

into account. Furthermore, the competition of intramolecular base-pairing with intermolecular pairing is usually neglected. Knowledge of the relevant rate constants in model cases would greatly enhance our understanding of

◄ **Fig. 3** Specificity and binding free energy. The *upper box* shows base-pairing schemes predicted for three cognate Mic/*omp* mRNA pairs (paired nucleotides are *white in black boxes*; *asterisks* indicate gaps). Calculated ΔG° values (25 °C) are given. Note that the experimentally confirmed base-pairing region for MicA/*ompA* RNA is shorter (see Fig. 1) than the one shown here. *Middle box*: Heterologous Mic/*omp* RNA combinations (entirely theoretical) with calculated ΔG° values. *Lower box*: The same analysis for an entirely unrelated sRNA, Spot 42. Predicted pairings are indicated by *white boxes*. Note that all interactions in the *two lower boxes* are purely hypothetical and only serve the discussion in Sect. 5.5. The values were calculated according to Hodas and Aalberts (2004)

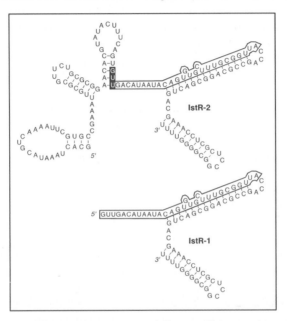

Fig. 4 Same target-complementary sequence, different efficiency: the IstR case. The *upper* and *lower secondary structures* represent IstR-2 and IstR-1, respectively, and are based on chemical and enzymatic probing (Darfeuille, unpublished results). The region of target (*tisAB* mRNA) complementarity is indicated by *boxes*. The *black box* highlights nucleotides that are sequestered in IstR-2

why a given sRNA binds one target but fails to bind another. For some (or many) sRNAs, proteins like Hfq may provide additional help by, for example, unfolding an inaccessible region, or perhaps by speeding up the (nonspecific) formation of an encounter complex, preceding base-pairing.

6
Concluding Remarks

We can anticipate that the prevalence of sRNAs in *E. coli* and its close relatives will be mirrored by many similar RNAs in many other bacteria. After all, it is unlikely that enterobacteria represent a special case. How many sRNAs will be found in *E. coli* alone is a matter of speculation, and estimates range from maybe 150 to more than 1000. Whatever number is closer to the truth, it is clear that searches have concentrated on certain expected classes and genomic locations. For example, *cis*-encoded antisense RNAs have only rarely been found (GadY, Sok variants, RdlR) and may have been overlooked owing to constraints and limitations in the search methods. Also, sRNAs encoded from within ORF regions (in either orientation) are difficult to identify. In particular, the ones in sense orientation cannot easily be distinguished from fortuitous breakdown/processing products of an mRNA (Vogel et al. 2003). Assignment of function is the critical step which is required to distinguish a bona fide sRNA from a short RNA segment present under a given set of conditions.

The quest for biological roles is ongoing in many laboratories. Though only relatively few sRNAs have been characterized in *E. coli*, stress responses so far score highest. In some other bacteria, regulation of virulence has been observed. What appears clear is that the vast majority of sRNAs are not essential for growth under standard laboratory conditions. However, their effects under special stress/growth conditions are likely to contribute to the significant fitness benefit that could explain their widespread use and conservation between related bacteria.

Similarly, the involvement of Hfq in sRNA function is a topic that requires attention. At this point, how this protein acts is not understood. The pleiotropic effects of *hfq* mutations make in vivo studies ambiguous, and in vitro studies are hampered by the difficulty to mimic the intracellular conditions in which Hfq normally acts (e.g., the presence of a vast number of competing RNAs at high concentration). Furthermore, a number of other putative helper proteins are present—their involvement in sRNA-mediated regulation has not been tested.

Finally, given that most sRNAs in *E. coli* appear to be antisense RNAs, some mechanistic questions arise. The complex dependence of the binding rate on the structures of the interacting RNAs and the elusive nature of specificity in light of many putative base-pairing possibilities are issues that deserve to be addressed. This is not only true for sRNAs. Regulatory RNAs in eukaryotes— miRNAs, siRNAs, and others—likely face the same challenges, such as specific target recognition and subsequent rapid pairing, and so comparative mechanistic studies may be enlightening in the future.

Acknowledgements We acknowledge support from grants by The Swedish Research Council, Wallenberg Consortium North, Human Frontiers Science Program Organization, and the EU-STREP program FOSRAK (Functions of small RNAs across kingdoms). F.D. is financed by a Marie Curie (EU) fellowship.

References

Aarons S, Abbas A, Adams C, Fenton A, O'Gara F (2000) A regulatory RNA (PrrB RNA) modulates expression of secondary metabolite genes in Pseudomonas fluorescens F113. J Bacteriol 182:3913–3919

Ali Azam T, Iwata A, Nishimura A, Ueda S, Ishihama A (1999) Growth phase-dependent variation in protein composition of the Escherichia coli nucleoid. J Bacteriol 181:6361–6370

Altuvia S (2004) Regulatory small RNAs: the key to coordinating global regulatory circuits. J Bacteriol 186:6679–6680

Altuvia S, Wagner EGH (2000) Switching on and off with RNA. Proc Natl Acad Sci USA 97:9824–9826

Altuvia S, Weinstein-Fischer D, Zhang A, Postow L, Storz G (1997) A small, stable RNA induced by oxidative stress: role as a pleiotropic regulator and antimutator. Cell 90:43–53

Altuvia S, Zhang A, Argaman L, Tiwari A, Storz G (1998) The Escherichia coli OxyS regulatory RNA represses fhlA translation by blocking ribosome binding. EMBO J 17:6069–6075

Andersen J, Delihas N (1990) micF RNA binds to the 5' end of ompF mRNA and to a protein from Escherichia coli. Biochemistry 29:9249–9256

Andersen J, Delihas N, Ikenaka K, Green PJ, Pines O, Ilercil O, Inouye M (1987) The isolation and characterization of RNA coded by the micF gene in Escherichia coli. Nucleic Acids Res 15:2089–2101

Andersen J, Forst SA, Zhao K, Inouye M, Delihas N (1989) The function of micF RNA. micF RNA is a major factor in the thermal regulation of OmpF protein in Escherichia coli. J Biol Chem 264:17961–17970

Argaman L, Hershberg R, Vogel J, Bejerano G, Wagner EGH, Margalit H, Altuvia S (2001) Novel small RNA-encoding genes in the intergenic regions of Escherichia coli. Curr Biol 11:941–950

Bartel DP (2004) MicroRNAs: genomics, biogenesis, mechanism, and function. Cell 116:281–297

Brescia CC, Mikulecky PJ, Feig AL, Sledjeski DD (2003) Identification of the Hfq-binding site on DsrA RNA: Hfq binds without altering DsrA secondary structure. RNA 9:33–43

Carpousis AJ (2003) Degradation of targeted mRNAs in Escherichia coli: regulation by a small antisense RNA. Genes Dev 17:2351–2355

Chen S, Lesnik EA, Hall TA, Sampath R, Griffey RH, Ecker DJ, Blyn LB (2002) A bioinformatics based approach to discover small RNA genes in the Escherichia coli genome. Biosystems 65:157–177

Chen S, Zhang A, Blyn LB, Storz G (2004) MicC, a second small-RNA regulator of Omp protein expression in Escherichia coli. J Bacteriol 186:6689–6697

Christensen SK, Pedersen K, Hansen FG, Gerdes K (2003) Toxin-antitoxin loci as stress-response-elements: ChpAK/MazF and ChpBK cleave translated RNAs and are counteracted by tmRNA. J Mol Biol 332:809–819

Deighan P, Free A, Dorman CJ (2000) A role for the Escherichia coli H-NS-like protein StpA in OmpF porin expression through modulation of micF RNA stability. Mol Microbiol 38:126–139

Geissmann TA, Touati D (2004) Hfq, a new chaperoning role: binding to messenger RNA determines access for small RNA regulator. EMBO J 23:396–405

Gerdes K, Gultyaev AP, Franch T, Pedersen K, Mikkelsen ND (1997) Antisense RNA-regulated programmed cell death. Annu Rev Genet 31:1–31

Gottesman S (2004) The small RNA regulators of Escherichia coli: roles and mechanisms. Annu Rev Microbiol:303–328

Gottesman S, Storz G, Rosenow C, Majdalani N, Repoila F, Wassarman KM (2001) Small RNA regulators of translation: mechanisms of action and approaches for identifying new small RNAs. Cold Spring Harbor Symp Quant Biol 66:353–362

He L, Hannon GJ (2004) MicroRNAs: small RNAs with a big role in gene regulation. Nat Rev Genet 5:522–531

Hengge-Aronis R (2002) Signal transduction and regulatory mechanisms involved in control of the sigma(S) (RpoS) subunit of RNA polymerase. Microbiol Mol Biol Rev 66:373–395

Hershberg R, Altuvia S, Margalit H (2003) A survey of small RNA-encoding genes in Escherichia coli. Nucleic Acids Res 31:1813–1820

Hindley J (1967) Fractionation of 32P-labelled ribonucleic acids on polyacrylamide gels and their characterization by fingerprinting. J Mol Biol 30:125–136

Hjalt TA, Wagner EGH (1995) Bulged-out nucleotides in an antisense RNA are required for rapid target RNA binding in vitro and inhibition in vivo. Nucleic Acids Res 23:580–587

Hodas NO, Aalberts DP (2004) Efficient computation of optimal oligo-RNA binding. Nucleic Acids Res 32:6636–6642

Huntzinger E, Boisset S, Saveanu C, Benito Y, Geissmann TA, Namane A, Kina G, Etienne J, Ehresmann B, Ehresmann C, Jacquier A, Vandenesch F, Romby P (2005) Staphylococcus aureus RNAIII and the endoribonuclease III coordinately regulate spa gene expression. EMBO J 24:824–835

Ikemura T, Dahlberg JE (1973) Small ribonucleic acids of Escherichia coli. II. Noncoordinate accumulation during stringent control. J Biol Chem 248:5033–5041

Jacob F, Monod J (1961) Genetic regulatory mechanisms in the synthesis of proteins. J Mol Biol 3:318–356

Kawamoto H, Morita T, Shimizu A, Inada T, Aiba H (2005) Implication of membrane localization of target mRNA in the action of a small RNA: mechanism of post-transcriptional regulation of glucose transporter in Escherichia coli. Genes Dev 19:328–338

Kawano M, Oshima T, Kasai H, Mori H (2002) Molecular characterization of long direct repeat (LDR) sequences expressing a stable mRNA encoding for a 35-amino-acid cell-killing peptide and a cis-encoded small antisense RNA in Escherichia coli. Mol Microbiol 45:333–349

Kolb FA, Engdahl HM, Slagter-Jäger JG, Ehresmann B, Ehresmann C, Westhof E, Wagner EGH, Romby P (2000) Progression of a loop-loop complex to a four-way junction is crucial for the activity of a regulatory antisense RNA. EMBO J 19:5905–5915

Kreikemeyer B, Boyle MD, Buttaro BA, Heinemann M, Podbielski A (2001) Group A streptococcal growth phase-associated virulence factor regulation by a novel operon (Fas) with homologies to two-component-type regulators requires a small RNA molecule. Mol Microbiol 39:392–406

Lease RA, Woodson SA (2004) Cycling of the Sm-like protein Hfq on the DsrA small regulatory RNA. J Mol Biol 344:1211–1223

Lease RA, Cusick ME, Belfort M (1998) Riboregulation in Escherichia coli: DsrA RNA acts by RNA:RNA interactions at multiple loci. Proc Natl Acad Sci USA 95:12456–12461

Lenz DH, Mok KC, Lilley BN, Kulkarni RV, Wingreen NS, Bassler BL (2004) The small RNA chaperone Hfq and multiple small RNAs control quorum sensing in Vibrio harveyi and Vibrio cholerae. Cell 118:69–82

Liu MY, Gui G, Wei B, Preston JF, 3rd, Oakford L, Yuksel U, Giedroc DP, Romeo T (1997) The RNA molecule CsrB binds to the global regulatory protein CsrA and antagonizes its activity in Escherichia coli. J Biol Chem 272:17 502–17 510

Majdalani N, Cunning C, Sledjeski D, Elliott T, Gottesman S (1998) DsrA RNA regulates translation of RpoS message by an anti-antisense mechanism, independent of its action as an antisilencer of transcription. Proc Natl Acad Sci USA 95:12462–12467

Majdalani N, Chen S, Murrow J, St John K, Gottesman S (2001) Regulation of RpoS by a novel small RNA: the characterization of RprA. Mol Microbiol 39:1382–1394

Majdalani N, Hernandez D, Gottesman S (2002) Regulation and mode of action of the second small RNA activator of RpoS translation, RprA. Mol Microbiol 46:813–826

Malmgren C, Engdahl HM, Romby P, Wagner EGH (1996) An antisense/target RNA duplex or a strong intramolecular RNA structure 5' of a translation initiation signal blocks ribosome binding: the case of plasmid R1. RNA 2:1022–1032

Mangold M, Siller M, Roppenser B, Vlaminckx BJ, Penfound TA, Klein R, Novak R, Novick RP, Charpentier E (2004) Synthesis of group A streptococcal virulence factors is controlled by a regulatory RNA molecule. Mol Microbiol 53:1515–1527

Massé E, Gottesman S (2002) A small RNA regulates the expression of genes involved in iron metabolism in Escherichia coli. Proc Natl Acad Sci USA 99:4620–4625

Massé E, Escorcia FE, Gottesman S (2003) Coupled degradation of a small regulatory RNA and its mRNA targets in Escherichia coli. Genes Dev 17:2374–2383

Moll I, Leitsch D, Steinhauser T, Bläsi U (2003) RNA chaperone activity of the Sm-like Hfq protein. EMBO Rep 4:284–289

Møller T, Franch T, Hojrup P, Keene DR, Bachinger HP, Brennan RG, Valentin-Hansen P (2002a) Hfq: a bacterial Sm-like protein that mediates RNA-RNA interaction. Mol Cell 9:23–30

Møller T, Franch T, Udesen C, Gerdes K, Valentin-Hansen P (2002b) Spot 42 RNA mediates discoordinate expression of the E. coli galactose operon. Genes Dev 16:1696–1706

Morfeldt E, Taylor D, von Gabain A, Arvidson S (1995) Activation of alpha-toxin translation in Staphylococcus aureus by the trans-encoded antisense RNA, RNAIII. EMBO J 14:4569–4577

Nordgren S, Slagter-Jäger JG, Wagner EGH (2001) Real time kinetic studies of the interaction between folded antisense and target RNAs using surface plasmon resonance. J Mol Biol 310:1125–1134

Nordström K, Wagner EGH (1994) Kinetic aspects of control of plasmid replication by antisense RNA. Trends Biochem Sci 19:294–300

Novick RP, Ross HF, Projan SJ, Kornblum J, Kreiswirth B, Moghazeh S (1993) Synthesis of staphylococcal virulence factors is controlled by a regulatory RNA molecule. EMBO J 12:3967–3975

Opdyke JA, Kang JG, Storz G (2004) GadY, a small-RNA regulator of acid response genes in Escherichia coli. J Bacteriol 186:6698–6705

Pedersen K, Gerdes K (1999) Multiple hok genes on the chromosome of Escherichia coli. Mol Microbiol 32:1090–1102

Repoila F, Gottesman S (2003) Temperature sensing by the dsrA promoter. J Bacteriol 185:6609–6614

Repoila F, Majdalani N, Gottesman S (2003) Small non-coding RNAs, co-ordinators of adaptation processes in Escherichia coli: the RpoS paradigm. Mol Microbiol 48:855–861

Rivas E, Klein RJ, Jones TA, Eddy SR (2001) Computational identification of noncoding RNAs in E. coli by comparative genomics. Curr Biol 11:1369–1373

Romeo T (1998) Global regulation by the small RNA-binding protein CsrA and the noncoding RNA molecule CsrB. Mol Microbiol 29:1321–1330

Schumacher MA, Pearson RF, Moller T, Valentin-Hansen P, Brennan RG (2002) Structures of the pleiotropic translational regulator Hfq and an Hfq-RNA complex: a bacterial Sm-like protein. EMBO J 21:3546–3556

Slagter-Jäger JG, Wagner EGH (2003) Loop swapping in an antisense RNA/target RNA pair changes directionality of helix progression. J Biol Chem 278:35 558–35 563

Sledjeski D, Gottesman S (1995) A small RNA acts as an antisilencer of the H-NS-silenced rcsA gene of Escherichia coli. Proc Natl Acad Sci USA 92:2003–2007

Sledjeski DD, Gupta A, Gottesman S (1996) The small RNA, DsrA, is essential for the low temperature expression of RpoS during exponential growth in Escherichia coli. EMBO J 15:3993–4000

Sledjeski DD, Whitman C, Zhang A (2001) Hfq is necessary for regulation by the untranslated RNA DsrA. J Bacteriol 183:1997–2005

Storz G, Opdyke JA, Zhang A (2004) Controlling mRNA stability and translation with small, noncoding RNAs. Curr Opin Microbiol 7:140–144

Stougaard P, Molin S, Nordström K (1981) RNAs involved in copy-number control and incompatibility of plasmid R1. Proc Natl Acad Sci USA 78:6008–6012

Suzuki K, Wang X, Weilbacher T, Pernestig AK, Melefors O, Georgellis D, Babitzke P, Romeo T (2002) Regulatory circuitry of the CsrA/CsrB and BarA/UvrY systems of Escherichia coli. J Bacteriol 184:5130–5140

Tetart F, Bouche JP (1992) Regulation of the expression of the cell-cycle gene ftsZ by DicF antisense RNA. Division does not require a fixed number of FtsZ molecules. Mol Microbiol 6:615–620

Tjaden B, Saxena RM, Stolyar S, Haynor DR, Kolker E, Rosenow C (2002) Transcriptome analysis of Escherichia coli using high-density oligonucleotide probe arrays. Nucleic Acids Res 30:3732–3738

Tomizawa J, Itoh T, Selzer G, Som T (1981) Inhibition of ColE1 RNA primer formation by a plasmid-specified small RNA. Proc Natl Acad Sci USA 78:1421–1425

Trotochaud AE, Wassarman KM (2004) 6S RNA function enhances long-term cell survival. J Bacteriol 186:4978–4985

Udekwu K, Darfeuille F, Vogel J, Reimegård J, Holmqvist E, Wagner EGH (2005) Hfq-dependent regulation of OmpA synthesis is mediated by an antisense RNA. Genes Dev (in press)

Urbanowski ML, Stauffer LT, Stauffer GV (2000) The gcvB gene encodes a small untranslated RNA involved in expression of the dipeptide and oligopeptide transport systems in Escherichia coli. Mol Microbiol 37:856–868

Valentin-Hansen P, Eriksen M, Udesen C (2004) The bacterial Sm-like protein Hfq: a key player in RNA transactions. Mol Microbiol 51:1525–1533

Valverde C, Lindell M, Wagner EG, Haas D (2004) A Rrepeated GGA motif is critical for the activity and stability of the riboregulator RsmY of Pseudomonas fluorescence. J Biol Chem 279:25066–25074

Vanderpool CK, Gottesman S (2004) Involvement of a novel transcriptional activator and small RNA in post-transcriptional regulation of the glucose phosphoenolpyruvate phosphotransferase system. Mol Microbiol 54:1076–1089

Vecerek B, Moll I, Afonyushkin T, Kaberdin V, Blasi U (2003) Interaction of the RNA chaperone Hfq with mRNAs: direct and indirect roles of Hfq in iron metabolism of Escherichia coli. Mol Microbiol 50:897–909

Vogel J, Wagner EGH (2005) Approaches to identify novel non-messenger RNAs in bacteria and to investigate their biological functions. I. RNA mining. In: Westhof E, Bindereif A, Schön A, Hartmann RK (eds) Handbook of biochemistry, vol 2. Wiley-VCH, Weinheim, Germany, pp 595–630

Vogel J, Bartels V, Tang TH, Churakov G, Slagter-Jäger JG, Hüttenhofer A, Wagner EGH (2003) RNomics in Escherichia coli detects new sRNA species and indicates parallel transcriptional output in bacteria. Nucleic Acids Res 31:6435–6443

Vogel J, Argaman L, Wagner EGH, Altuvia S (2004) The small RNA IstR inhibits synthesis of an SOS-induced toxic peptide. Curr Biol 14:2271–2276

Vytvytska O, Moll I, Kaberdin VR, von Gabain A, Blasi U (2000) Hfq (HF1) stimulates ompA mRNA decay by interfering with ribosome binding. Genes Dev 14:1109–1118

Wagner EGH, Flärdh K (2002) Antisense RNAs everywhere? Trends Genet 18:223–226

Wagner EGH, Vogel J (2003) Noncoding RNAs encoded by bacterial chromosomes. In: Barciszewski J, Erdmann V (eds) Noncoding RNAs. Landes Bioscience, Georgetown, TX, USA, pp 243–259

Wagner EGH, Vogel J (2005) Approaches to identify novel non-messenger RNAs in bacteria and to investigate their biological functions. II. Functional analysis of identified non-mRNA. In: Westhof E, Bindereif A, Schön A, Hartmann RK (eds) Handbook of biochemistry, vol 2. Wiley-VCH, Weinheim, Germany, pp 614–654

Wagner EGH, Altuvia S, Romby P (2002) Antisense RNAs in bacteria and their genetic elements. Adv Genet 46:361–398

Waldsich C, Grossberger R, Schroeder R (2002) RNA chaperone StpA loosens interactions of the tertiary structure in the td group I intron in vivo. Genes Dev 16:2300–2312

Wassarman KM (2002) Small RNAs in bacteria: diverse regulators of gene expression in response to environmental changes. Cell 109:141–144

Wassarman KM, Storz G (2000) 6S RNA regulates E. coli RNA polymerase activity. Cell 101:613–623

Wassarman KM, Zhang A, Storz G (1999) Small RNAs in Escherichia coli. Trends Microbiol 7:37–45

Wassarman KM, Repoila F, Rosenow C, Storz G, Gottesman S (2001) Identification of novel small RNAs using comparative genomics and microarrays. Genes Dev 15:1637–1651

Weilbacher T, Suzuki K, Dubey AK, Wang X, Gudapaty S, Morozov I, Baker CS, Georgellis D, Babitzke P, Romeo T (2003) A novel sRNA component of the carbon storage regulatory system of Escherichia coli. Mol Microbiol 48:657–670

Wilderman PJ, Sowa NA, FitzGerald DJ, FitzGerald PC, Gottesman S, Ochsner UA, Vasil ML (2004) Identification of tandem duplicate regulatory small RNAs in Pseudomonas aeruginosa involved in iron homeostasis. Proc Natl Acad Sci USA 101:9792–9797

Zhang A, Altuvia S, Tiwari A, Argaman L, Hengge-Aronis R, Storz G (1998) The OxyS regulatory RNA represses rpoS translation and binds the Hfq (HF-I) protein. EMBO J 17:6061–6068

Zhang A, Wassarman KM, Ortega J, Steven AC, Storz G (2002) The Sm-like Hfq protein increases OxyS RNA interaction with target mRNAs. Mol Cell 9:11–22

Zhang A, Wassarman KM, Rosenow C, Tjaden BC, Storz G, Gottesman S (2003) Global analysis of small RNA and mRNA targets of Hfq. Mol Microbiol 50:1111–1124

Nucleic Acids and Molecular Biology, Vol. 17
Wolfgang Nellen, Christian Hammann (Eds.)
Small RNAs
© Springer-Verlag Berlin Heidelberg 2005

Small Nucleolar RNAs:
Identification, Structure, and Function

Fredrik Söderbom

Department of Molecular Biology, Biomedical Center,
Swedish University of Agricultural Sciences, Box 590, 75124 Uppsala, Sweden
fredde@xray.bmc.uu.se

Abstract The revelation in the last few years of a large number of new noncoding RNAs (ncRNAs) has revolutionized our view of how gene regulation works. This is to a large extent due to the recent advances in computational as well as experimental methodology, combined with an increasing number of sequenced genomes which have had an enormous impact on the quest for new small ncRNAs. These RNAs have a function per se and are not merely intermediates in the transfer of information from genes to proteins. Instead they have turned out to be involved in the regulation of many complex biological processes, including stress response, cell differentiation, and even in control of diseases. This chapter briefly describes some of the methods used to identify and isolate different classes of ncRNAs in the size range 50–500 nt. One of these, small nucleolar RNAs, will be discussed in detail.

1
Introduction

Historically, RNA was considered to be nucleic acid carrying the information required to produce proteins. The different RNA molecules that were involved in the transfer of genetic information from DNA to protein were transfer RNA (tRNA), ribosomal RNA (rRNA), and messenger RNA (mRNA)—no other classes of RNA were necessary. This view of RNA as being solely an intermediate in the information transfer has now changed dramatically. Experimental and computational screens for noncoding RNA (ncRNA) in the last few years have revealed a plethora of ncRNAs in all kingdoms of life (reviewed in Eddy 2001; Hüttenhofer et al. 2002; Storz 2002). These systematic searches uncovered an unexpected abundance and functional diversity of ncRNAs. Besides being required for protein synthesis, ncRNAs are now also known to be involved in a number of fundamental biological processes, e.g., RNA modification and processing, regulation of development by post-transcriptional control of gene expression, splicing, telomere maintenance, targeting proteins for membrane insertion, and chromatin structure formation, just to mention a few (Eddy 2001 and references therein). The great number of newly identified ncRNAs have not only expanded our view of the versatility of RNA, but have also given us new insights into the functions of nucleic acid sequences that until recently were considered as "junk", i.e., introns and intergenic DNA. Many ncRNAs are derived from introns that have been spliced out and subsequently processed by ribonucleases to form mature functional RNAs, whereas others are encoded from DNA regions that previously were considered silent, e.g., not transcribed (Eddy 2001; Mattick 2001; Mattick and Gagen 2001).

In this review, experimental and computational methods for isolation and identification of ncRNAs in the size range of 50–500 nt will be discussed. Furthermore, one large class of ncRNA, small nucleolar RNA (snoRNA), of which many new members have recently been discovered, will be described in detail.

2
Experimental Isolation of ncRNAs

The different experimental methods that have been used to isolate small ncRNAs can be divided into two main approaches: (1) specific isolation, where the aim is to isolate certain known classes of RNAs, and (2) general unbiased procedures, where RNA is isolated on the basis of criteria such as size.

2.1
Specific Isolation

Some classes of ncRNAs are commonly associated with distinct sets of proteins forming ribonucleoprotein (RNP) particles. Antibodies that specifically recognize proteins in RNPs can be used to isolate the associated RNA. This approach was employed to isolate small nuclear RNAs (snRNAs) involved in splicing. Patients with the autoimmune disease systemic lupus erythematosus produce antibodies against highly conserved cellular components. These components were purified and turned out to be RNPs containing snRNAs U1, U2, U4, U5, and U6 (Lerner and Steitz 1981). Antibodies have also been used to isolate another class of small ncRNA, namely snoRNA. This class of RNA is involved in modification and processing of certain RNAs (see later). Cavaillé and coworkers isolated snoRNA from rat brain by immunoprecipitation using antibodies against fibrillarin, a protein that specifically binds to box C/D snoRNAs. To further enrich this family of snoRNA, the RNA was converted into complementary DNA (cDNA) using reverse transcription PCR primers designed to only amplify RNA molecules containing sequence elements specific for box C/D snoRNAs (Cavaillé et al. 2001). A similar approach has been employed to isolate the other main family of snoRNAs, box H/ACA snoRNAs from human cells (Kiss et al. 2004).

Immunoprecipitation has also been used to isolate snoRNAs from *Trypanosoma brucei* (Dunbar et al. 2000b) as well as sno-like RNAs (sRNAs) from archaea (Omer et al. 2000). Another approach is to isolate subcellular fractions in which the desired RNA is accumulated, e.g., the nucleolytic fraction for snoRNAs (Kiss-Laszlo et al. 1996; Zhou et al. 2004). Since the majority of snoRNAs in vertebrates are derived from intronic sequences, further enrichment of box C/D snoRNAs from HeLa cells was achieved by constructing cDNA libraries in such way that only RNA molecules that had been processed from longer RNA transcripts could be amplified and cloned (Kiss-Laszlo et al. 1996).

2.2
General Isolation

Several large-scale screens for RNAs in the size range 50–500 nt have recently identified a great number of new ncRNAs in different organisms from all kingdoms of life (Hüttenhofer et al. 2002; Vogel et al. 2003). This approach, to generate cDNA libraries from size-fractionated small RNAs, was coined experimental RNomics and has been used to identify hundreds of new ncRNAs in different model organisms, e.g., bacteria (Vogel et al. 2003), archaea (Tang et al. 2002; Tang et al. 2005), *Arabidopsis* (Marker et al. 2002), *Drosophila* (Yuan et al. 2003), and mouse (Hüttenhofer et al. 2001). Briefly, RNA is isolated and size-fractionated by polyacrylamide gel electrophoresis. RNA sized 50–500 nt is extracted from gel slices and the 3′ ends of the RNA molecules are C-tailed by CTP and poly(A) polymerase followed by reverse transcription using a poly d(G) oligonucleotide. After second-strand cDNA synthesis, linkers are added and the cDNA is cloned into vectors. To enrich for cDNAs representing novel RNAs, the cDNA clones (or PCR-amplified inserts) derived from abundant known ncRNAs, e.g., rRNA, tRNA, and snRNA, are excluded by hybridization screening. The remaining novel ncRNA candidates are sequenced and analyzed (Hüttenhofer et al. 2004). Using a modified version of the RNomics approach (Fig. 1), a number of novel ncRNAs were identified in *Dictyostelium discoideum*. The major difference from the method just described was that the generated cDNA library represented *full-length* small RNAs. This was achieved by adding an additional step, where T4 RNA ligase was used to ligate an RNA oligonucleotide to the 5′ end of the RNA. To allow ligation to both primary and processed transcripts, the RNA was first treated with tobacco acid pyrophosphatase to remove cap structures/convert 5′-triphosphates of primary transcripts to monophosphates (Aspegren et al. 2004).

RNAs frequently identified in these systematic screens are signal recognition particle (SRP) RNA, RNase P RNA, and snRNAs. Furthermore, entirely novel classes of RNAs, with unknown functions, have been discovered, e.g., class I and class II RNAs in *Dictyostelium* (Aspegren et al. 2004). One of the major surprises arising from the RNomic approach was the large number of snoRNAs present in eukaryotes and archaea. This has set off an increasing interest concerning snoRNAs and many new findings have recently been reported, some of which will be discussed in detail in this chapter.

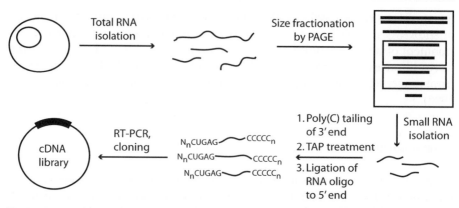

Fig. 1 Construction of a complementary DNA (*cDNA*) library representing full-length RNA. Total RNA was isolated and size-fractionated by polyacrylamide gel electrophoresis (*PAGE*), and RNA of desired length was extracted from gel slices. The RNA was C-tailed and treated with tobacco acid pyrophosphatase (*TAP*) to remove cap structures from primary transcripts before an RNA oligonucleotide was ligated to the 5′ end. The tagged RNA was converted to cDNA by reverse transcription PCR (*RT-PCR*) using primers specific for the 5′ and 3′ tagged ends. The cDNA products were cloned into plasmid vectors (Aspegren et al. 2004)

3
Computational Approaches to Find ncRNAs

Many different computational approaches have been used in attempts to identify ncRNA genes (Eddy 2002; Gräf et al., this volume). Classes of ncRNAs usually have structural features in common, e.g., stem-loops, but searches based solely on secondary structure motifs do not seem to be useful as a general method to find ncRNA genes (Rivas and Eddy 2000). The best results have been obtained when genomes of closely related organisms have been sequenced, thus permitting comparative genome analysis. This allows for comparison of sequence features that are conserved between related organisms and thereby greatly improves the chances of finding true ncRNAs (Coventry et al. 2004; McCutcheon and Eddy 2003; Rivas and Eddy 2001; Washietl et al. 2005). Examples of conserved features are intergenic location and secondary structures with compensatory base changes that conserve the structure. Different computational screens for ncRNAs in *Escherichia coli* have successfully used comparative genomics alone or in combination with search methods such as promoter/terminator predictions and microarray analysis to identify new ncRNAs (Argaman et al. 2001; Rivas et al. 2001; Wassarman et al. 2001). Other general bioinformatic searches were based on the observation that ncRNA genes in A/T-rich genomes of hyperthermophilic archaea, i.e., *Methanococcus jannashii* and *Pyrococcus furiosus*, have a higher G/C content compared with that of the surrounding DNA (Klein et al. 2002; Schattner

2002). A similar approach applied to the A/T-rich genome of the protist *Dictyostelium* is presently being performed (Larsson and Söderbom, unpublished results).

Specific classes of RNAs can be identified on the basis of conserved structural and/or sequence motifs, e.g., tRNAs (tRNAscan-SE; Lowe and Eddy 1997), SRP RNAs (SRPscan; Regalia et al. 2002), and RNase P RNA (Li and Altman 2004). Many box C/D snoRNAs have been successfully predicted by computational methods on the basis of their conserved sequence and structural elements as well as their guide sequences that bind to the complementary target sequences via antisense interaction, e.g., in yeasts (Lowe and Eddy 1999; Qu et al. 1999), plants (Barneche et al. 2001; Brown et al. 2001; Qu et al. 2001; see also Plant snoRNA data base, Brown et al. 2003b), and archaea (Gaspin et al. 2000; Omer et al. 2000). While box C/D snoRNAs have been relatively easy to predict, computational searches for box H/ACA snoRNAs have proven much harder owing to shorter and less well-conserved sequence motifs. However, two publications recently reported computational searches (and verified expression) for box H/ACA snoRNAs in yeasts (Schattner et al. 2004) and *Drosophila* (Huang et al. 2004).

Many ncRNAs are listed in different databases available on the web. One of these is Rfam, http://www.sanger.ac.uk/Software/Rfam/ (Griffiths-Jones et al. 2003, 2005), where sequences of many different classes of ncRNAs from a variety of organisms are available. Furthermore, Rfam can be used to annotate sequences (including complete genomes) by searching for homologs to known ncRNAs.

4
Small Nucleolar RNAs

The recent isolation of a large number of snoRNAs from different organisms has helped us to understand the functions of these modification guide RNAs. This class of RNA, approximately 60–150 nt in length, is involved in modification and processing of precursor rRNA (pre-rRNA) and has recently been shown to be responsible for guiding modification of other RNAs as well. The following sections will focus on structural and functional aspects of modification guide snoRNAs, their association with proteins as well as their genomic organization. The different snoRNAs involved in pre-rRNA cleavage and the snoRNA domain of telomerase RNA will be only briefly addressed in this review.

snoRNAs were first assigned a function in pre-rRNA processing, although it was noted early that this group of RNAs contained sequences complementary to rRNA. This observation suggested additional roles, e.g., in rRNA folding and rRNA-protein assembly. It was also observed that modified nucleotides within rRNA were situated in regions complementary to sequences in snoRNAs, raising the suggestion that snoRNAs had a role in rRNA modification (Bachellerie et al. 1995a; Steitz and Tycowski 1995). This later suggestion turned out to hold true, and it is now known that the majority of the snoRNAs function as guides for modification of rRNA nucleotides. Recently, other targets for these guide RNAs have emerged, e.g., snRNAs involved in splicing, tRNAs, and possibly mRNAs (see later).

The snoRNAs can be divided into two main families on the basis of short conserved sequence motifs; box C/D snoRNAs, which mediate 2'-O-ribose methylation, and box H/ACA snoRNAs, which guide pseudouridylation of RNA (Fig. 2; Balakin et al. 1996). Common to both families, they contain one or more sequence(s) that can base-pair with their target sequence(s) and

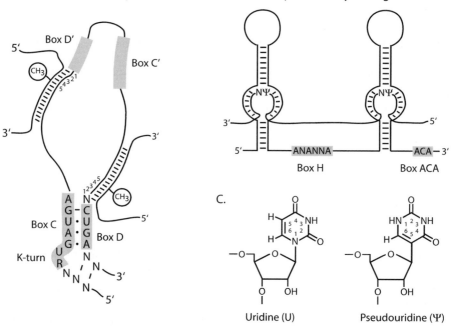

A. Box C/D methylation guide RNA

B. Box H/ACA pseudouridylation guide RNA

C.

Uridine (U)

Pseudouridine (Ψ)

Fig. 2 a Schematic structures of box C/D guide RNAs. The conserved C and D sequences are displayed in *gray boxes* together with the evolutionarily conserved kink-turn (*K-turn*) secondary structure. The less conserved C' and D' motifs are depicted as *gray boxes*. Target RNAs, e.g. ribosomal RNAs, base-pair to complementary small nucleolar RNA (*snoRNA*) sequences positioned immediately 5' of the D and/or D' boxes. The nucleotide to be methylated in the target RNA is base-paired to the fifth nucleotide upstream of the D or D' box. **b** Schematic structure of the box H/ACA guide RNA family. The conserved sequence motifs, H and ACA, are boxed. The target RNAs base-pair to internal loops within the guide RNA forming two short helix structures, separated by two unpaired nucleotides of which one is the uridine that will be converted to a pseudouridine (Ψ). **c** The conversion of uridines to pseudouridines is guided by box H/ACA RNAs

thereby specifically mark the nucleotide to be modified. In addition, each family is associated with a specific set of core proteins, forming small nucleolar RNP (snoRNP) particles. Some of the snoRNAs of both families are required for cleavage of pre-rRNA (reviewed in Henras et al. 2004b; Venema and Tollervey 1999). Another type of snoRNA that is involved in pre-rRNA cleavage is the RNA component of endoribonuclease MRP which is evolutionary related to RNase P (Chamberlain et al..1998; Lygerou et al. 1996). Human telomerase RNA (hTR) contains an H/ACA domain at its 3′ half which is required for hTR accumulation, 3′ end processing, and telomerase activity (Mitchell et al. 1999). Recently snoRNA-like RNAs were identified in archaea, suggesting that these RNA molecules may be of ancient evolutionary origin (Gaspin et al. 2000; Omer et al. 2000; Tang et al. 2002).

4.1
Box C/D snoRNAs

Modification box C/D snoRNAs guide 2′-O-ribose methylation of RNA (Fig. 2a). They share conserved sequence motifs important for function, protein interaction, and biogenesis. The box C motif (5′-RUGAUGA-3′) is located close to the 5′ end, whereas the D motif (5′-CUGA-3′) is situated at the 3′ end of the RNA (Bachellerie and Cavaillé 1998; Tyc and Steitz 1989). In addition, two less conserved box C and D motifs, C′ and D′, are often present within the RNA (Kiss-Laszlo et al. 1998; Tycowski et al. 1996a). The box C/D snoRNAs contain sequences upstream of box D and/or box D′ that are complementary to their target RNA. Upon antisense interaction between these sequences, the snoRNA can guide methylation of a target nucleotide, i.e., the methyltransferase associated with the snoRNA performs the actual methylation. The importance of the conserved sequence and structural motifs for correct snoRNA formation and function has been analyzed in a number of studies.

The termini of many of the box C/D snoRNAs identified appear to form a short (approximately 4–5-bp) stem-domain, keeping the 5′ and 3′ ends together. This terminal stem, flanked by the C and D boxes, constitutes a terminal core motif (Xia et al. 1997) which brings the conserved sequence motifs together, allowing them to work in concert for correct processing and accumulation of box C/D snoRNAs (Caffarelli et al. 1996; Cavaillé and Bachellerie 1996; Huang et al. 1992; Li and Fournier 1992; Xia et al. 1997). However, this stem is not present in all snoRNAs (Aspegren et al. 2004; Bachellerie et al. 1995a; Dunbar et al. 2000a) and can be compensated for by external and/or internal base-pairing (Darzacq and Kiss 2000; Villa et al. 2000). This highlights the importance of base-pairing as a means to keep boxes C and D in close vicinity for efficient assembly of snoRNA-specific proteins. Furthermore, the conserved box C and D sequences have been demonstrated to be crucial for localization of the snoRNAs to the nucleolus (Lange et al. 1998;

Narayanan et al. 1999b; Samarsky et al. 1998). In addition, it was recently shown that a functional box D sequence is required for efficient transcription and 3′-end formation of box C/D snoRNAs (Morlando et al. 2004). Furthermore, another form of box C/D RNAs is present in the hyperthermophilic archaeon *Pyrococcus furiosus*. Here, the box C/D RNAs are present in the rare form of RNA circles (Starostina et al. 2004).

The box C/D snoRNAs are very specific and guide 2′-O-ribose methylation of distinct target nucleotides. Each snoRNA is usually responsible for the modification of one unique nucleotide. How is this remarkable specificity achieved, and how does the snoRNA "know" which target nucleotide to act upon?

Each snoRNA has a sequence (10–21 nt) situated immediately 5′ of the D and/or D′ box(es) which is complementary to its target RNA (Bachellerie et al. 1995b; Jarmolowski et al. 1990; Kiss-Laszlo et al. 1996). The complementary snoRNA sequence and the target sequence base-pair (Beltrame and Tollervey 1995; Liang and Fournier 1995) and the nucleotide to be modified is always located within the complementary sequence of the target RNA (Kiss-Laszlo et al. 1996; Nicoloso et al. 1996). This still does not tell us what singles out one specific nucleotide for modification. It turns out that the D, and sometimes the D′, motif has a specific function. It determines which nucleotide will be modified. The D and D′ boxes act as rulers, measuring the distance to the target nucleotide (Fig. 2a). It is always the nucleotide paired to the fifth nucleotide upstream of these motifs that is modified (Cavaillé et al. 1996; Kiss-Laszlo et al. 1996; Nicoloso et al. 1996).

The function and specificity of box C/D snoRNAs were experimentally demonstrated in vivo by expressing artificial box C/D snoRNAs designed to target alternative rRNA sequences; new and specific 2′-O-ribose methylations were introduced by exchanging the antisense elements of certain box C/D snoRNAs for sequences that were predicted to interact with nucleotides that normally are not methylated (Cavaillé et al. 1996). Also, depleting the cell of specific snoRNAs abolishes/suppresses the 2′-O-methylation of the targeted nucleotides (Kiss-Laszlo et al. 1996; Tycowski et al. 1996b). Further evidence for the role of box C/D snoRNAs in directing site-specific methylation was obtained in vitro by demonstrating that purified box C/D snoRNPs from *Saccharomyces cerevisiae* could direct specific 2′-O-methylation. This reaction is dependent on correct RNA–RNA interaction between the snoRNA and its target sequence (Galardi et al. 2002). In vitro reconstitution of archaeal box C/D sRNP guided methylation has also been achieved (Omer et al. 2002; Tran et al. 2003).

4.2
Box H/ACA snoRNAs

The other main members of the snoRNA class are the box H/ACA snoRNAs. These RNAs guide the isomerization of specific target uridines to pseudouridines (Fig. 2c). The box H/ACA snoRNAs are characterized by conserved sequence as well as structural motifs (Fig. 2b). The common structure (hairpin–hinge–hairpin–tail) consists of an ACA motif at the 3′ end and two stems-loops linked by a hinge (H) sequence. The conserved H box comprises the sequence A*NANNA* (where *N* stands for any nucleotide) and the ACA box is situated 3 nt from the 3′ end of the RNA (Balakin et al. 1996; Ganot et al. 1997b). Both boxes, ACA and H (Balakin et al. 1996; Ganot et al. 1997b), as well as each of the two hairpin structures (Balakin et al. 1996; Bortolin et al. 1999), were shown to be required for accumulation of the RNAs. In addition, both conserved sequence motifs, H and ACA, constitute nucleolar localization elements (Lange et al. 1999; Narayanan et al. 1999a).

The ability of H/ACA snoRNAs to direct site-specific pseudouridylation in rRNA was revealed by correlating the disruption of certain snoRNA genes with the failure to form specific pseudouridines. These modifications could be rescued by reintroducing the corresponding snoRNA genes (Ganot et al. 1997a; Ni et al. 1997). Furthermore, pseudouridylation of specific RNA sequences was demonstrated in vitro using purified box H/ACA snoRNPs (Wang et al. 2002).

In analogy with the box C/D snoRNA family, the box H/ACA snoRNAs direct site-specific modification by interacting with complementary RNA sequences that contain the uridine to be modified. The box H/ACA snoRNA sequences that interact with the target RNA are situated at opposite strands of internal loops (called pseudouridylation pockets) within one or both of the conserved stem-loop structures. Thus, each snoRNA can thereby guide the modification of one and sometimes two uridines. In each pseudouridylation pocket, the snoRNA forms two short helical structures of 3–10 nt each with the target sequence, separated by two unpaired nucleotides (Ganot et al. 1997a; Ni et al. 1997). The first unpaired nucleotide (5′–3′ direction of the target RNA) is the uridine that will be converted to pseudouridine (Fig. 2b,c). Another common theme shared between the two classes of snoRNAs is the function of conserved sequences as "measuring devices." While D (and/or D′) motifs in box C/D snoRNAs define which nucleotide is to be methylated, the distance between the target uridine and the H or ACA boxes, usually 14–16 nt, is critical for modification of the correct uridine (Ganot et al. 1997a; Ni et al. 1997; Wang et al. 2002).

By expressing variants of a human box H/ACA snoRNA in yeasts, it was demonstrated that both hairpin structures are required for pseudouridylation even if the box H/ACA snoRNA only guides the modification of one uridine, i.e., when only one hairpin is involved in target interaction (Bortolin et al. 1999). In spite of this, small RNAs related to pseudouridylation guide RNAs, which do not conform to the typical hairpin–hinge–hairpin–tail structure, have been identified. In archaea, trypanosomes, and *Euglena gracilis*, box H/ACA-like snoRNAs have been reported to consist of a single hairpin structure (Liang et al. 2001, 2004; Liang et al. 2002b; Rozhdestvensky et al. 2003;

Russell et al. 2004; Tang et al. 2002). Furthermore, in archaea, box H/ACA-like RNAs can consist of up to three stem-loops (Rozhdestvensky et al. 2003; Tang et al. 2002) whereas Cajal body-specific guide RNAs have been shown to exist as two tandemly repeated fused box H/ACA domains with four stem-loop structures (Kiss et al. 2002) or are even composed of box H/ACA and C/D domains fused together within the same RNA (Darzacq et al. 2002).

5
Small Cajal-Body-Specific RNAs

Recently, higher eukaryotes were shown to encode a certain subclass of modification guide RNAs. These were denoted small Cajal body-specific RNAs (scaRNAs) since they reside in the Cajal bodies where they target snRNAs (involved in splicing) for modification. Cajal bodies (CBs) are nucleoplasmic organelles in which a number of nuclear factors are colocalized, e.g., spliceosomal small nuclear RNPs and snoRNPs. Recent data suggest that CBs are involved in maturation and assembly of nuclear RNPs (Gerbi et al. 2003; Ogg and Lamond 2002, and references therein). It appears as if canonical snoRNAs (at least box C/D snoRNAs), after their synthesis in the nucleoplasm, transit through CBs, where maturation and assembly into snoRNP particles probably takes place, before accumulating in the nucleolus (Boulon et al. 2004; Narayanan et al. 1999b; Richard et al. 2003; Samarsky et al. 1998; Verheggen et al. 2002).

Guide sequences identified in scaRNAs so far are predicted to direct pseudouridylation and 2′-O-methylation of specific nucleotides within U1, U2, U4, and U5 snRNAs (Darzacq et al. 2002; Jady et al. 2003; Jady and Kiss 2001; Kiss et al. 2002, 2004). These guide RNAs are specifically located to CBs and are very similar to snoRNAs, harboring the conserved sequence and structural motifs associated with box C/D and H/ACA snoRNAs. Frequently, the scaRNAs are composed of both C/D and H/ACA core motifs, but recently a number of consensus box H/ACA RNAs have been isolated (Darzacq et al. 2002; Kiss et al. 2004; Richard et al. 2003).

What makes the scaRNAs stay in the CBs, rather than continuing to the nucleolus? In the scaRNAs that contain box H/ACA motifs, an evolutionarily conserved sequence, the CB box (CAB), was identified in the two terminal loops of the 5′ and 3′ hairpins. This sequence motif is responsible for localization of box H/ACA scaRNAs to CBs (Richard et al. 2003). The CAB, UGAG, where AG is highly conserved among the box H/ACA scaRNAs identified, may constitute a binding site for protein(s) that could retain the H/ACA scaRNA within CBs. If the CABs are mutated, the scaRNAs accumulate in the nucleolus, like canonical snoRNAs (Richard et al. 2003).

For the box C/D scaRNAs identified (Darzacq et al. 2002), the elements responsible for CB localization remain to be determined.

6
Protein Association and Function

The snoRNAs do not function alone but must be associated with specific proteins, to form nucleolar RNP particles that constitute the functional modification entities in the cell. Each class of snoRNAs binds to its specific set of proteins which are required for both biogenesis and function. It is important to remember that the RNA moiety solely acts as a guide and provides the specificity, while proteins carry the enzymatic activity to introduce modifications.

6.1
Box C/D snoRNPs

Four essential core proteins are associated with eukaryotic box C/D snoRNAs (and most likely box C/D scaRNAs): fibrillarin (animals and plants)/Nop1p (yeasts), Nop56p, Nop58p, and 15.5-kD protein (vertebrates)/Snu13p (yeasts) (Gautier et al. 1997; Lafontaine and Tollervey 1999, 2000; Lyman et al. 1999; Schimmang et al. 1989; Watkins et al. 2000; Wu et al. 1998). Three orthologs of the eukaryotic proteins are present in archaea box C/D sRNPs, i.e., L7 (15.5-kD protein), a single Nop56/58p (Nop56p/Nop58p), and fibrillarin. These proteins interact with box C/D sRNAs to direct 2'-O-methylation (Kuhn et al. 2002; Omer et al. 2002; Tran et al. 2003).

The C and D motifs can be folded into a stem-internal asymmetrical loop-stem structure that keeps the hallmark sequence motif close together (Fig. 2a). This motif is conserved among the box C/D snoRNAs (Vidovic et al. 2000) and is also found in box H/ACA sRNAs in archaea (Rozhdestvensky et al. 2003). This structural motif, termed the kink-turn (K-turn) (Klein et al. 2001), is also present in the 5' stem-loop of U4 snRNA as well as in other ncRNAs and seems to constitute a conserved RNA binding motif (Klein et al. 2001; Vidovic et al. 2000). Binding of the Snu13p/15.5-kD protein to the box C/D motif/K-turn motif is probably necessary for subsequent binding of the other core box C/D snoRNP-associated proteins. This protein–RNA interaction possibly generates binding sites for the remaining snoRNP proteins by changing the structure of the snoRNA (Szewczak et al. 2002; Vidovic et al. 2000; Watkins et al. 2000, 2002), since the crystal structures of the 15.5-kD protein binding to the K-turn of the 5' stem-loop of U4 snRNA display major structural rearrangements within the RNA (Vidovic et al. 2000).

Nop56p and Nop58p share extensive sequence similarities, and Nop58p has been reported to be required for snoRNA accumulation (Gautier et al. 1997; Lafontaine and Tollervey 1999, 2000; Lyman et al. 1999; Wu et al. 1998). Cross-linking studies revealed that Nop58 interacts with the C motif, and Nop56 with the C' sequence (Cahill et al. 2002).

Several lines of evidence pinpoint fibrillarin/Nop1p as the protein responsible for the methyltransferase reaction guided by box C/D RNAs. The crystal structure of the fibrillarin ortholog from the archaeon *Methanococcus jannaschii* revealed that the C-terminal domain is structurally homologous to the catalytic domain found in many methyltransferases (Wang et al. 2000). Furthermore, mutations in yeast Nop1p, affecting rRNA methylation, were shown to be located in the conserved region of the putative methyltransferase domain (Tollervey et al. 1993; Wang et al. 2000). Moreover, Galardi and coworkers showed that purified box C/D snoRNPs could direct site-specific methylation of rRNA, and that methylation was abolished when wild-type Nop1p was replaced with one carrying a mutation affecting rRNA methylation in vivo (Galardi et al. 2002; Tollervey et al. 1993; and see before). It was also shown, by cross-linking analysis, that fibrillarin interacts with boxes D' and D (and C') which position the putative 2'-O-methyltransferase in the vicinity of the target nucleotide to be methylated (Cahill et al. 2002).

6.2
Box H/ACA snoRNPs

A different set of specific proteins is associated with the other major family of snoRNAs, box H/ACA snoRNAs. These snoRNPs consist of four essential major proteins; Nhp2p (Henras et al. 1998; Kolodrubetz and Burgum 1991; Watkins et al. 1998), Nop10p (Henras et al. 1998), Cbf5p (NAP57 in rats, Meier and Blobel 1994; Dyskerin in humans, Heiss et al. 1998; Jiang et al. 1993; Lafontaine et al. 1998; Watkins et al. 1998; Yang et al. 2000), and Gar1p (Balakin et al. 1996; Ganot et al. 1997b; Girard et al. 1992; Watkins et al. 1998). Candidate archaeal homologs for the eukaryotic box H/ACA snoRNA-associated proteins have been identified. One of these, ribosomal protein L7 (homologous to Nhp2p), was shown to interact with archaeal box H/ACA RNA (Rozhdestvensky et al. 2003; Watanabe and Gray 2000; Watkins et al. 1998).

Depletion of any of the protein components leads to impaired global rRNA pseudouridylation (Bousquet-Antonelli et al. 1997; Henras et al. 1998; Lafontaine et al. 1998), whereas gene knockouts of individual box H/ACA snoRNAs abolish the isomerization of specific target uridines to pseudouridines (see later). In most cases, the lack of uridine isomerase activity caused by reduced levels of the snoRNP core proteins is probably an indirect effect, since the accumulation of the components forming the box H/ACA snoRNPs seems to be tightly coregulated. The levels of the associated proteins Cbf5p, Gar1p, and Nop10p are dependent on each other since depletion of one of these proteins inhibits accumulation of the others. By contrast, Nhp2p levels are only marginally decreased; however, depletion of Nhp2p causes reduced levels of the other three protein components (Henras et al. 1998, 2004a; Lafontaine et al. 1998). Furthermore, depletion of Nhp2p, Nop10p, or Cbf5p leads to reduced stability of H/ACA snoRNAs (Henras et al. 1998; Lafontaine et al. 1998; Watkins et al. 1998). Conversely, Gar1p can be depleted without major effects on the levels of the pseudouridine guide RNAs. Instead, Gar1p can directly bind to H/ACA RNAs in vitro (Bagni and Lapeyre 1998) and has been suggested to play a role in the interaction between the box H/ACA snoRNPs and pre-rRNA (Bousquet-Antonelli et al. 1997).

Recently, a model was presented of the assembly of mammalian H/ACA RNPs based on in vitro assays, e.g., protein–protein and protein–RNA interactions. In this model, the four proteins associate in a given order to form a complex before binding to the RNA. Nop10p associates with NAP57, followed by binding of Nhp2p, while Gar1p independently associates with NAP57. When this protein complex is formed, it associates with the H/ACA RNAs to create functional H/ACA RNPs (Wang and Meier 2004).

Which protein is responsible for pseudouridylation is still not entirely clear, although compelling evidence points at Cbf5p/NAP57 as the active enzyme. Purified box H/ACA snoRNPs can pseudouridylate their target RNAs in vitro, and the only protein component in the RNA–protein complex that shows similarity to a pseudouridine synthase is Cbf5p/NAP57 (Koonin 1996; Wang et al. 2002). Furthermore, when highly conserved amino acids in the predicted catalytic pseudouridine synthase domain of Cbf5p were altered, pseudouridylation of rRNA was reduced or abolished (Zebarjadian et al. 1999). In addition, pseudouridylation target uridines can be UV-cross-linked to NAP57 (Wang and Meier 2004).

7
Genomic Organization

Eukaryotic organisms show a great variation in the way snoRNA genes are organized in their genomes. In vertebrates, all known modification guide RNAs are derived from introns, whereas in plants and lower eukaryotes, the majority of the snoRNAs are transcribed form their own genes, either as single guide RNAs or as part of a polycistronic transcript that contains several snoRNAs (Fig. 3).

The majority of mammalian modification guide snoRNAs are derived from introns, containing a single snoRNA, located in genes involved in ribosome biogenesis or protein synthesis. However, several of them are encoded within genes with no apparent protein coding capacity (Cavaillé et al. 2001, 2002; Kiss et al. 2004; Pelczar and Filipowicz 1998; Smith and Steitz 1998; Tycowski et al. 1996a; Vitali et al. 2003). This raises the possibility that the only function of these genes is to encode pre-RNAs with introns that will be processed into functional snoRNAs (Kiss et al. 2004; Tycowski et al. 1996a). Furthermore, several snoRNA genes in mammals reside in imprinted loci, and some snoRNAs are tissue-specifically expressed (Cavaillé et al. 2000, 2002). The organization of snoRNA genes in *Drosophila* follows the same rule as for vertebrates, e.g., they are situated in introns. However, some *Drosophila* box

Independently transcribed snoRNA genes

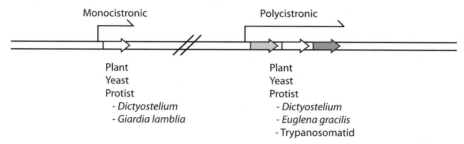

Monocistronic	Polycistronic

Plant
Yeast
Protist
  - *Dictyostelium*
  - *Giardia lamblia*

Plant
Yeast
Protist
  - *Dictyostelium*
  - *Euglena gracilis*
  - Trypanosomatid

Intron-encoded snoRNA genes

Intron of protein-coding gene (Monocistronic)

Intron of protein-coding gene (Polycistronic)

Intron of non-coding RNA gene (Monocistronic)

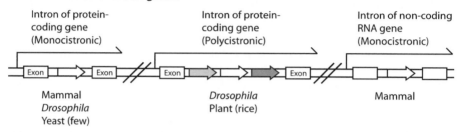

Mammal
Drosophila
Yeast (few)

Drosophila
Plant (rice)

Mammal

Fig. 3 Schematic genome sequence illustrating the organization of snoRNA genes in different organisms. The snoRNA genes (*open arrows*) can be transcribed from their own promoters either as single genes or as gene clusters. The snoRNAs can also be derived from introns (hence cotranscribed with the precursor messenger RNA). Mammalian snoRNAs are exclusively located within introns (one snoRNA/intron), whereas clusters of snoRNAs within the same intron can be found in rice and *Drosophila*

H/ACA guide RNAs are polycistronic, i.e., several snoRNA genes are clustered within the same intron (Huang et al. 2004; Yuan et al. 2003).

In plants, snoRNAs are generally present in polycistronic clusters. In *Arabidopsis* and maize, the great majority of the snoRNAs are processed from nonintronic polycistronic transcripts (Barneche et al. 2001; Brown et al. 2001; Leader et al. 1997; Qu et al. 2001), whereas in rice the clusters are often intronic (Chen et al. 2003; Liang et al. 2002a). Interestingly, a so far unique snoRNA organization was reported in plants where some box C/D snoRNAs are expressed together with tRNAs from dicistronic tRNA–snoRNA genes (Kruszka et al. 2003).

Different types of genomic organization of snoRNAs are found in yeasts. The majority are transcribed independently from intergenic regions, i.e., in between protein coding genes, but several are found in polycistronic snoRNA gene clusters, and a few are generated from introns (Chanfreau et al. 1998; Lowe and Eddy 1999; Qu et al. 1999; Samarsky and Fournier 1999; Schattner et al. 2004). In *Dictyostelium*, snoRNAs appear to be encoded by single genes, or transcribed as polycistronic units that can contain both box C/D and H/ACA snoRNAs (Aspegren et al. 2004). Transcripts with representatives from both families of snoRNAs are also present in plants (Chen et al. 2003; Leader et al. 1997) and lower eukaryotes. No intronic snoRNAs have so far been found in *Dictyostelium*, although this organism branched out after plants but before metazoans and fungi, placing it close to the base of metazoan evolution (Aspegren et al. 2004; Baldauf et al. 2000; Glockner et al. 2002). The organization of snoRNA genes in polycistronic clusters carrying both C/D and H/ACA RNAs is also found in early diverging eukaryotes, e.g., *E. gracilis* and trypanosomatids (Dunbar et al. 2000a; Liang et al. 2004; Roberts et al. 1998; Russell et al. 2004; Xu et al. 2001), while in *Giardia lamblia*, one of the earliest offset of eukaryotes, the majority of the snoRNA-like RNAs identified are transcribed independently from their own genes (Yang et al. 2005).

8
Modification Guide RNAs—Targets and Function

The main function of modification guide RNPs appears to be related to pseudouridylation and 2'-O-methylation in RNA, but additional roles may also be of importance. The guide RNPs and their role in modification are evolutionarily conserved from archaea to mammals: each guide RNA usually targets one, but sometimes several, specific nucleotide(s) for modification (in yeasts up to three 2'-O-methylations, Kiss-Laszlo et al. 1996, and as many as four pseudouridylations, Schattner et al. 2004).

In vertebrate rRNAs more than 100 nucleotides are 2'-O-methylated, and 90–100 uridines are converted to pseudouridines. The number in plants has been estimated to exceed 120 of each modification, and the rRNA in yeasts contains about 44–55 of each type (Brown et al. 2003a; Maden 1990; Ofengand and Bakin 1997). The modified nucleotides are commonly located in ribosomal regions which are conserved and functionally important, e.g., the peptidyl transferase center (Decatur and Fournier 2002). One could easily imagine that this conservation would imply that each modification is, if not essential, at least very important for the function of the ribosome and, hence, the viability of the cell. Surprisingly, modifications are dispensable, and their absence does not result in any major effect on cell growth. This has been demonstrated in yeasts by constructing gene knockouts where one or several snoRNA genes has/have been disrupted, which prevented modification of their target nucleotide(s) (Balakin et al. 1996; King et al. 2003; Kiss-Laszlo et al. 1996; Lowe and Eddy 1999; Qu et al. 1999; Samarsky et al. 1995).

Small nuclear spliceosomal RNAs, U1, U2, U4, U5, and U6, are also targets for modification. In vertebrates and plants this is mainly directed by box C/D and box H/ACA guide RNAs. Significant complementarity to sequences surrounding many of the modified nucleotides has been identified and in a few cases it has been demonstrated that certain guide RNAs are required for specific modifications (Darzacq et al. 2002; Ganot et al. 1999; Hüttenhofer et al. 2001; Jady and Kiss 2001; Kiss et al. 2002; Marker et al. 2002; Tycowski et al. 1998; Vitali et al. 2003; Yu et al. 2001; Zhao et al. 2002; Zhou et al. 2002).

The sites of modifications in spliceosomal RNAs follow the same rule as for rRNA, i.e., they are predominantly clustered in regions of functional importance (Gu et al. 1996). However, the significance of guide RNAs directing modification of U snRNAs for cell growth and function is still enigmatic (Zhou et al. 2002).

Other targets for guide RNAs include the spliced leader (SL) RNA (reviewed in Hastings 2005) in trypanosomatids where the conversion of a uridine to a pseudouridine is guided by a box H/ACA guide RNA (Liang et al. 2002b), and an archaeal precursor tRNA (pre-trRNA)^{Trp} which harbors an intron containing a box C/D sRNA which can guide 2'-O-methylation of the pre-tRNA (Bortolin et al. 2003; Singh et al. 2004). Furthermore, effort in the last few years to experimentally isolate box C/D and H/ACA RNAs from different organisms has revealed a great number of orphans, i.e., putative guide RNAs with no obvious antisense complementarity to any of the known classes of target RNAs (Cavaillé et al. 2000, 2002; Hüttenhofer et al. 2001; Jady and Kiss 2000; Kiss et al. 2004; Marker et al. 2002; Vitali et al. 2003; Yuan et al. 2003). Interestingly, some of these orphan snoRNAs display tissue-specific expression.

9
Conclusion and Future Perspective

In spite of the effort in the last few years to isolate, identify, and characterize the majority of box C/D and box H/ACA guide RNAs (and their associated proteins) from a number of different organisms, several questions still remain unanswered. Which are the targets for the orphan snoRNAs that lack complementarities to RNAs normally modified by guide RNPs? What are the functions of the snoRNAs that are expressed only in specific tissues, e.g., the brain? How does the CAB mediate retention of scaRNAs within CBs, which protein(s) interact with this sequence and what motif(s) are responsible for accumulation of scaRNAs that lack the CAB? Moreover, the importance of the canonical modification guide box C/D and box H/ACA snoRNAs is not completely understood. The nucleotide modifications they direct appear to be dispensable for cell growth since disruption of snoRNA genes in most cases does not affect cell viability. However, when pseudouridylation or 2'-O-methylation of rRNA was blocked globally in S. cerevisiae, by introducing point mutations in the catalytic sites of Cbf5p (the putative pseudouridine synthase) or Nop1p (the putative methylase), the cells showed slow growth or died (Tollervey et al. 1993; Zebarjadian et al. 1999). This suggests that many of the modified nucleotides in the rRNA function in a synergistic way and that individual modifications may have subtler roles, e.g., in fine-tuning RNA–RNA and RNA–protein interactions. However, it is important to keep in mind that preventing the modifying action of Cbf5p and Nop1p could possibly affect the function of other RNAs as well. These enzymes are part of box C/D and box H/ACA guide RNPs, respectively, and may not only influence the modification status of rRNA but also that of other target RNAs, such as spliceosomal RNAs and possibly mRNAs recognized by orphan guide RNPs. However, internal modification of U snRNAs in *S. cerevisiae* has been suggested to be guide-RNA independent (Ma et al. 2003). What are the functions of all these modifications? The conversion of 2′-OH to a 2′-*O*-methyl can stabilize RNA structures (Davis 1998). This may be the reason why the number of methylation guide sRNA genes detected in archaeal genomes increases with the optimal growth temperature of the organism (Dennis et al. 2001). Pseudouridines have also been shown to have a stabilizing effect on RNA structure by enhancing local base stacking and stabilizing base-pair interactions, thereby acting as a molecular glue to maintain required RNA conformations (Davis 1998; Ofengand 2002). Guide RNAs per se may also have a direct function on RNA structure by acting as chaperones. Since both box C/D and box H/ACA guide RNAs interact with their target RNAs by base-pairing, they could aid in orchestrating the correct folding of RNA (Bachellerie et al. 1995b; Maxwell and Fournier 1995).

The recent advances in the understanding of the biosynthesis and function of modification guide RNAs and their associated proteins have increased the knowledge concerning these ncRNAs to a great extent. Many of the unanswered questions raised herein will most likely be solved in the near future but considering the recent exciting new findings, more surprises will most certainly hide around the corner.

Acknowledgements I thank Gerhart Wagner and Andrea Hinas for critical reading of the manuscript and Andrea Hinas for preparing the figures. This work was supported by the Swedish Research Council and Wallenberg Consortium North.

References

Argaman L, Hershberg R, Vogel J, Bejerano G, Wagner EGH, Margalit H, Altuvia S (2001) Novel small RNA-encoding genes in the intergenic regions of Escherichia coli. Curr Biol 11:941–950

Aspegren A, Hinas A, Larsson P, Larsson A, Söderbom F (2004) Novel non-coding RNAs in Dictyostelium discoideum and their expression during development. Nucleic Acids Res 32:4646–4656

Bachellerie JP, Cavaillé J (1998) Small nucleolar RNAs guide the ribose methylations of eukaryotic rRNAs. In: Grosjean H, Benne R (eds) Modification and editing of RNA. American Society for Microbiology, Washington, DC, pp 255–272

Bachellerie JP, Michot B, Nicoloso M, Balakin A, Ni J, Fournier MJ (1995a) Antisense snoRNAs: a family of nucleolar RNAs with long complementarities to rRNA. Trends Biochem Sci 20:261–264

Bachellerie JP, Nicoloso M, Qu LH, Michot B, Caizergues-Ferrer M, Cavaillé J, Renalier MH (1995b) Novel intron-encoded small nucleolar RNAs with long sequence complementarities to mature rRNAs involved in ribosome biogenesis. Biochem Cell Biol 73:835–843

Bagni C, Lapeyre B (1998) Gar1p binds to the small nucleolar RNAs snR10 and snR30 in vitro through a nontypical RNA binding element. J Biol Chem 273:10868–10873

Balakin AG, Smith L, Fournier MJ (1996) The RNA world of the nucleolus: two major families of small RNAs defined by different box elements with related functions. Cell 86:823–834

Baldauf SL, Roger AJ, Wenk-Siefert I, Doolittle WF (2000) A kingdom-level phylogeny of eukaryotes based on combined protein data. Science 290:972–977

Barneche F, Gaspin C, Guyot R, Echeverria M (2001) Identification of 66 box C/D snoRNAs in Arabidopsis thaliana: extensive gene duplications generated multiple isoforms predicting new ribosomal RNA 2'-O-methylation sites. J Mol Biol 311:57–73

Beltrame M, Tollervey D (1995) Base pairing between U3 and the pre-ribosomal RNA is required for 18S rRNA synthesis. EMBO J 14:4350–4356

Bortolin ML, Bachellerie JP, Clouet-d'Orval B (2003) In vitro RNP assembly and methylation guide activity of an unusual box C/D RNA, cis-acting archaeal pre-tRNA(Trp). Nucleic Acids Res 31:6524–6535

Bortolin ML, Ganot P, Kiss T (1999) Elements essential for accumulation and function of small nucleolar RNAs directing site-specific pseudouridylation of ribosomal RNAs. EMBO J 18:457–469

Boulon S, Verheggen C, Jady BE, Girard C, Pescia C, Paul C, Ospina JK, Kiss T, Matera AG, Bordonne R, Bertrand E (2004) PHAX and CRM1 are required sequentially to transport U3 snoRNA to nucleoli. Mol Cell 16:777–787

Bousquet-Antonelli C, Henry Y, G'Elugne JP, Caizergues-Ferrer M, Kiss T (1997) A small nucleolar RNP protein is required for pseudouridylation of eukaryotic ribosomal RNAs. EMBO J 16:4770–4776

Brown JW, Clark GP, Leader DJ, Simpson CG, Lowe T (2001) Multiple snoRNA gene clusters from Arabidopsis. RNA 7:1817–1832

Brown JW, Echeverria M, Qu LH (2003a) Plant snoRNAs: functional evolution and new modes of gene expression. Trends Plant Sci 8:42–49

Brown JW, Echeverria M, Qu LH, Lowe TM, Bachellerie JP, Hüttenhofer A, Kastenmayer JP, Green PJ, Shaw P, Marshall DF (2003b) Plant snoRNA database. Nucleic Acids Res 31:432–435

Caffarelli E, Fatica A, Prislei S, De Gregorio E, Fragapane P, Bozzoni I (1996) Processing of the intron-encoded U16 and U18 snoRNAs: the conserved C and D boxes control both the processing reaction and the stability of the mature snoRNA. EMBO J 15:1121–1131

Cahill NM, Friend K, Speckmann W, Li ZH, Terns RM, Terns MP, Steitz JA (2002) Site-specific cross-linking analyses reveal an asymmetric protein distribution for a box C/D snoRNP. EMBO J 21:3816–3828

Cavaillé J, Bachellerie JP (1996) Processing of fibrillarin-associated snoRNAs from pre-mRNA introns: an exonucleolytic process exclusively directed by the common stem-box terminal structure. Biochimie 78:443–456

Cavaillé J, Nicoloso M, Bachellerie JP (1996) Targeted ribose methylation of RNA in vivo directed by tailored antisense RNA guides. Nature 383:732–735

Cavaillé J, Buiting K, Kiefmann M, Lalande M, Brannan CI, Horsthemke B, Bachellerie JP, Brosius J, Hüttenhofer A (2000) Identification of brain-specific and imprinted small nucleolar RNA genes exhibiting an unusual genomic organization. Proc Natl Acad Sci USA 97:14311–14316

Cavaillé J, Vitali P, Basyuk E, Hüttenhofer A, Bachellerie JP (2001) A novel brain-specific box C/D small nucleolar RNA processed from tandemly repeated introns of a noncoding RNA gene in rats. J Biol Chem 276:26374–26383

Cavaillé J, Seitz H, Paulsen M, Ferguson-Smith AC, Bachellerie JP (2002) Identification of tandemly-repeated C/D snoRNA genes at the imprinted human 14q32 domain reminiscent of those at the Prader-Willi/Angelman syndrome region. Hum Mol Genet 11:1527–1538

Chamberlain JR, Lee Y, Lane WS, Engelke DR (1998) Purification and characterization of the nuclear RNase P holoenzyme complex reveals extensive subunit overlap with RNase MRP. Genes Dev 12:1678–1690

Chanfreau G, Rotondo G, Legrain P, Jacquier A (1998) Processing of a dicistronic small nucleolar RNA precursor by the RNA endonuclease Rnt1. EMBO J 17:3726–3737

Chen CL, Liang D, Zhou H, Zhuo M, Chen YQ, Qu LH (2003) The high diversity of snoRNAs in plants: identification and comparative study of 120 snoRNA genes from Oryza sativa. Nucleic Acids Res 31:2601–2613

Coventry A, Kleitman DJ, Berger B (2004) MSARI: multiple sequence alignments for statistical detection of RNA secondary structure. Proc Natl Acad Sci USA 101:12102–12107

Darzacq X, Kiss T (2000) Processing of intron-encoded box C/D small nucleolar RNAs lacking a 5′,3′-terminal stem structure. Mol Cell Biol 20:4522–4531

Darzacq X, Jady BE, Verheggen C, Kiss AM, Bertrand E, Kiss T (2002) Cajal body-specific small nuclear RNAs: a novel class of 2′-O-methylation and pseudouridylation guide RNAs. EMBO J 21:2746–2756

Davis DR (1998) Biophysical and conformational properties of modified nucleosides in RNA (nuclear magnetic resonance studies). In: Grosjean H, Benne R (eds) Modification and editing of RNA. American Society for Microbiology, Washington, DC, pp 85–102

Decatur WA, Fournier MJ (2002) rRNA modifications and ribosome function. Trends Biochem Sci 27:344–351

Dennis PP, Omer A, Lowe T (2001) A guided tour: small RNA function in Archaea. Mol Microbiol 40:509–519

Dunbar DA, Chen AA, Wormsley S, Baserga SJ (2000a) The genes for small nucleolar RNAs in Trypanosoma brucei are organized in clusters and are transcribed as a polycistronic RNA. Nucleic Acids Res 28:2855–2861

Dunbar DA, Wormsley S, Lowe TM, Baserga SJ (2000b) Fibrillarin-associated box C/D small nucleolar RNAs in Trypanosoma brucei. Sequence conservation and implications for 2′-O-ribose methylation of rRNA. J Biol Chem 275:14767–14776

Eddy SR (2001) Non-coding RNA genes and the modern RNA world. Nat Rev Genet 2:919–929

Eddy SR (2002) Computational genomics of noncoding RNA genes. Cell 109:137–140

Galardi S, Fatica A, Bachi A, Scaloni A, Presutti C, Bozzoni I (2002) Purified box C/D snoRNPs are able to reproduce site-specific 2′-O-methylation of target RNA in vitro. Mol Cell Biol 22:6663–6668

Ganot P, Bortolin ML, Kiss T (1997a) Site-specific pseudouridine formation in preribosomal RNA is guided by small nucleolar RNAs. Cell 89:799–809

Ganot P, Caizergues-Ferrer M, Kiss T (1997b) The family of box ACA small nucleolar RNAs is defined by an evolutionarily conserved secondary structure and ubiquitous sequence elements essential for RNA accumulation. Genes Dev 11:941–956

Ganot P, Jady BE, Bortolin ML, Darzacq X, Kiss T (1999) Nucleolar factors direct the 2'-O-ribose methylation and pseudouridylation of U6 spliceosomal RNA. Mol Cell Biol 19:6906–6917

Gaspin C, Cavaillé J, Erauso G, Bachellerie JP (2000) Archaeal homologs of eukaryotic methylation guide small nucleolar RNAs: lessons from the Pyrococcus genomes. J Mol Biol 297:895–906

Gautier T, Berges T, Tollervey D, Hurt E (1997) Nucleolar KKE/D repeat proteins Nop56p and Nop58p interact with Nop1p and are required for ribosome biogenesis. Mol Cell Biol 17:7088–7098

Gerbi SA, Borovjagin AV, Lange TS (2003) The nucleolus: a site of ribonucleoprotein maturation. Curr Opin Cell Biol 15:318–325

Girard JP, Lehtonen H, Caizergues-Ferrer M, Amalric F, Tollervey D, Lapeyre B (1992) GAR1 is an essential small nucleolar RNP protein required for pre-rRNA processing in yeast. EMBO J 11:673–682

Glockner G, Eichinger L, Szafranski K, Pachebat JA, Bankier AT, Dear PH, Lehmann R, Baumgart C, Parra G, Abril JF, Guigo R, Kumpf K, Tunggal B, Cox E, Quail MA, Platzer M, Rosenthal A, Noegel AA (2002) Sequence and analysis of chromosome 2 of Dictyostelium discoideum. Nature 418:79–85

Griffiths-Jones S, Bateman A, Marshall M, Khanna A, Eddy SR (2003) Rfam: an RNA family database. Nucleic Acids Res 31:439–441

Griffiths-Jones S, Moxon S, Marshall M, Khanna A, Eddy SR, Bateman A (2005) Rfam: annotating non-coding RNAs in complete genomes. Nucleic Acids Res 33 Database Issue: D121–124

Gu J, Patton JR, Shimba S, Reddy R (1996) Localization of modified nucleotides in Schizosaccharomyces pombe spliceosomal small nuclear RNAs: modified nucleotides are clustered in functionally important regions. RNA 2:909–918

Hastings KEM (2005) SL trans-splicing: easy come or easy go? Trends in Genetics 21:240–247

Heiss NS, Knight SW, Vulliamy TJ, Klauck SM, Wiemann S, Mason PJ, Poustka A, Dokal I (1998) X-linked dyskeratosis congenita is caused by mutations in a highly conserved gene with putative nucleolar functions. Nat Genet 19:32–38

Henras A, Henry Y, Bousquet-Antonelli C, Noaillac-Depeyre J, Gelugne JP, Caizergues-Ferrer M (1998) Nhp2p and Nop10p are essential for the function of H/ACA snoRNPs. EMBO J 17:7078–7090

Henras AK, Capeyrou R, Henry Y, Caizergues-Ferrer M (2004a) Cbf5p, the putative pseudouridine synthase of H/ACA-type snoRNPs, can form a complex with Gar1p and Nop10p in absence of Nhp2p and box H/ACA snoRNAs. RNA 10:1704–1712

Henras AK, Dez C, Henry Y (2004b) RNA structure and function in C/D and H/ACA s(no)RNPs. Curr Opin Struct Biol 14:335–343

Huang GM, Jarmolowski A, Struck JC, Fournier MJ (1992) Accumulation of U14 small nuclear RNA in Saccharomyces cerevisiae requires box C, box D, and a 5', 3' terminal stem. Mol Cell Biol 12:4456–4463

Huang ZP, Zhou H, Liang D, Qu LH (2004) Different expression strategy: multiple intronic gene clusters of box H/ACA snoRNA in Drosophila melanogaster. J Mol Biol 341:669–683

Hüttenhofer A, Kiefmann M, Meier-Ewert S, O'Brien J, Lehrach H, Bachellerie JP, Brosius J (2001) RNomics: an experimental approach that identifies 201 candidates for novel, small, non-messenger RNAs in mouse. EMBO J 20:2943–2953

Hüttenhofer A, Brosius J, Bachellerie JP (2002) RNomics: identification and function of small, non-messenger RNAs. Curr Opin Chem Biol 6:835–843

Hüttenhofer A, Cavaillé J, Bachellerie JP (2004) In: Gott JM (ed) Methods in molecular biology, vol 265. Experimental RNomics: a global approach to identifying small nuclear RNAs and their targets in different model organisms. Humana, Totowa, NJ, pp 409–428

Jady BE, Kiss T (2000) Characterisation of the U83 and U84 small nucleolar RNAs: two novel 2′-O-ribose methylation guide RNAs that lack complementarities to ribosomal RNAs. Nucl Acids Res 28:1348–1354

Jady BE, Kiss T (2001) A small nucleolar guide RNA functions both in 2′-O-ribose methylation and pseudouridylation of the U5 spliceosomal RNA. EMBO J 20:541–551

Jady BE, Darzacq X, Tucker KE, Matera AG, Bertrand E, Kiss T (2003) Modification of Sm small nuclear RNAs occurs in the nucleoplasmic Cajal body following import from the cytoplasm. EMBO J 22:1878–1888

Jarmolowski A, Zagorski J, Li HV, Fournier MJ (1990) Identification of essential elements in U14 RNA of Saccharomyces cerevisiae. EMBO J 9:4503–4509

Jiang W, Middleton K, Yoon HJ, Fouquet C, Carbon J (1993) An essential yeast protein, CBF5p, binds in vitro to centromeres and microtubules. Mol Cell Biol 13:4884–4893

King TH, Liu B, McCully RR, Fournier MJ (2003) Ribosome structure and activity are altered in cells lacking snoRNPs that form pseudouridines in the peptidyl transferase center. Mol Cell 11:425–435

Kiss AM, Jady BE, Darzacq X, Verheggen C, Bertrand E, Kiss T (2002) A Cajal body-specific pseudouridylation guide RNA is composed of two box H/ACA snoRNA-like domains. Nucleic Acids Res 30:4643–4649

Kiss AM, Jady BE, Bertrand E, Kiss T (2004) Human box H/ACA pseudouridylation guide RNA machinery. Mol Cell Biol 24:5797–5807

Kiss-Laszlo Z, Henry Y, Bachellerie JP, Caizergues-Ferrer M, Kiss T (1996) Site-specific ribose methylation of preribosomal RNA: a novel function for small nucleolar RNAs. Cell 85:1077–1088

Kiss-Laszlo Z, Henry Y, Kiss T (1998) Sequence and structural elements of methylation guide snoRNAs essential for site-specific ribose methylation of pre-rRNA. EMBO J 17:797–807

Klein DJ, Schmeing TM, Moore PB, Steitz TA (2001) The kink-turn: a new RNA secondary structure motif. EMBO J 20:4214–4221

Klein RJ, Misulovin Z, Eddy SR (2002) Noncoding RNA genes identified in AT-rich hyperthermophiles. Proc Natl Acad Sci USA 99:7542–7547

Kolodrubetz D, Burgum A (1991) Sequence and genetic analysis of NHP2: a moderately abundant high mobility group-like nuclear protein with an essential function in Saccharomyces cerevisiae. Yeast 7:79–90

Koonin EV (1996) Pseudouridine synthases: four families of enzymes containing a putative uridine-binding motif also conserved in dUTPases and dCTP deaminases. Nucleic Acids Res 24:2411–2415

Kruszka K, Barneche F, Guyot R, Ailhas J, Meneau I, Schiffer S, Marchfelder A, Echeverria M (2003) Plant dicistronic tRNA-snoRNA genes: a new mode of expression of the small nucleolar RNAs processed by RNase Z. EMBO J 22:621–632

Kuhn JF, Tran EJ, Maxwell ES (2002) Archaeal ribosomal protein L7 is a functional homolog of the eukaryotic 15.5kD/Snu13p snoRNP core protein. Nucleic Acids Res 30:931–941

Lafontaine DL, Tollervey D (1999) Nop58p is a common component of the box C+D snoRNPs that is required for snoRNA stability. RNA 5:455–467

Lafontaine DL, Tollervey D (2000) Synthesis and assembly of the box C+D small nucleolar RNPs. Mol Cell Biol 20:2650–2659

Lafontaine DL, Bousquet-Antonelli C, Henry Y, Caizergues-Ferrer M, Tollervey D (1998) The box H + ACA snoRNAs carry Cbf5p, the putative rRNA pseudouridine synthase. Genes Dev 12:527–537

Lange TS, Borovjagin A, Maxwell ES, Gerbi SA (1998) Conserved boxes C and D are essential nucleolar localization elements of U14 and U8 snoRNAs. EMBO J 17:3176–3187

Lange TS, Ezrokhi M, Amaldi F, Gerbi SA (1999) Box H and box ACA are nucleolar localization elements of U17 small nucleolar RNA. Mol Biol Cell 10:3877–3890

Leader DJ, Clark GP, Watters J, Beven AF, Shaw PJ, Brown JW (1997) Clusters of multiple different small nucleolar RNA genes in plants are expressed as and processed from polycistronic pre-snoRNAs. EMBO J 16:5742–5751

Lerner MR, Steitz JA (1981) Snurps and scyrps. Cell 25:298–300

Li D, Fournier MJ (1992) U14 function in Saccharomyces cerevisiae can be provided by large deletion variants of yeast U14 and hybrid mouse-yeast U14 RNAs. EMBO J 11:683–689

Li Y, Altman S (2004) In search of RNase P RNA from microbial genomes. RNA 10:1533–1540

Liang D, Zhou H, Zhang P, Chen YQ, Chen X, Chen CL, Qu LH (2002a) A novel gene organization: intronic snoRNA gene clusters from Oryza sativa. Nucleic Acids Res 30:3262–3272

Liang WQ, Fournier MJ (1995) U14 base-pairs with 18S rRNA: a novel snoRNA interaction required for rRNA processing. Genes Dev 9:2433–2443

Liang XH, Liu L, Michaeli S (2001) Identification of the first trypanosome H/ACA RNA that guides pseudouridine formation on rRNA. J Biol Chem 276:40313–40318

Liang XH, Xu YX, Michaeli S (2002b) The spliced leader-associated RNA is a trypanosome-specific sn(o) RNA that has the potential to guide pseudouridine formation on the SL RNA. RNA 8:237–246

Liang XH, Ochaion A, Xu YX, Liu Q, Michaeli S (2004) Small nucleolar RNA clusters in trypanosomatid Leptomonas collosoma. Genome organization, expression studies, and the potential role of sequences present upstream from the first repeated cluster. J Biol Chem 279:5100–5109

Lowe TM, Eddy SR (1997) tRNAscan-SE: a program for improved detection of transfer RNA genes in genomic sequence. Nucleic Acids Res 25:955–964

Lowe TM, Eddy SR (1999) A computational screen for methylation guide snoRNAs in yeast. Science 283:1168–1171

Lygerou Z, Allmang C, Tollervey D, Seraphin B (1996) Accurate processing of a eukaryotic precursor ribosomal RNA by ribonuclease MRP in vitro. Science 272:268–270

Lyman SK, Gerace L, Baserga SJ (1999) Human Nop5/Nop58 is a component common to the box C/D small nucleolar ribonucleoproteins. RNA 5:1597–1604

Ma X, Zhao X, Yu YT (2003) Pseudouridylation (Psi) of U2 snRNA in S. cerevisiae is catalyzed by an RNA-independent mechanism. EMBO J 22:1889–1897

Maden BE (1990) The numerous modified nucleotides in eukaryotic ribosomal RNA. Prog Nucleic Acid Res Mol Biol 39:241–303

Marker C, Zemann A, Terhorst T, Kiefmann M, Kastenmayer JP, Green P, Bachellerie JP, Brosius J, Hüttenhofer A (2002) Experimental RNomics. Identification of 140 Candidates for Small Non- Messenger RNAs in the Plant Arabidopsis thaliana. Curr Biol 12:2002–2013

Mattick JS (2001) Non-coding RNAs: the architects of eukaryotic complexity. EMBO Rep 2:986–991

Mattick JS, Gagen MJ (2001) The evolution of controlled multitasked gene networks: the role of introns and other noncoding RNAs in the development of complex organisms. Mol Biol Evol 18:1611–1630

Maxwell ES, Fournier MJ (1995) The small nucleolar RNAs. Annu Rev Biochem 64:897–934

McCutcheon JP, Eddy SR (2003) Computational identification of non-coding RNAs in Saccharomyces cerevisiae by comparative genomics. Nucleic Acids Res 31:4119–4128

Meier UT, Blobel G (1994) NAP57, a mammalian nucleolar protein with a putative homolog in yeast and bacteria. J Cell Biol 127:1505–1514

Mitchell JR, Cheng J, Collins K (1999) A box H/ACA small nucleolar RNA-like domain at the human telomerase RNA 3′ end. Mol Cell Biol 19:567–576

Morlando M, Ballarino M, Greco P, Caffarelli E, Dichtl B, Bozzoni I (2004) Coupling between snoRNP assembly and 3′ processing controls box C/D snoRNA biosynthesis in yeast. EMBO J 23:2392–2401

Narayanan A, Lukowiak A, Jady BE, Dragon F, Kiss T, Terns RM, Terns MP (1999a) Nucleolar localization signals of box H/ACA small nucleolar RNAs. EMBO J 18:5120–5130

Narayanan A, Speckmann W, Terns R, Terns MP (1999b) Role of the box C/D motif in localization of small nucleolar RNAs to coiled bodies and nucleoli. Mol Biol Cell 10:2131–2147

Ni J, Tien AL, Fournier MJ (1997) Small nucleolar RNAs direct site-specific synthesis of pseudouridine in ribosomal RNA. Cell 89:565–573

Nicoloso M, Qu LH, Michot B, Bachellerie JP (1996) Intron-encoded, antisense small nucleolar RNAs: the characterization of nine novel species points to their direct role as guides for the 2′-O-ribose methylation of rRNAs. J Mol Biol 260:178–195

Ofengand J (2002) Ribosomal RNA pseudouridines and pseudouridine synthases. FEBS Lett 514:17–25

Ofengand J, Bakin A (1997) Mapping to nucleotide resolution of pseudouridine residues in large subunit ribosomal RNAs from representative eukaryotes, prokaryotes, archaebacteria, mitochondria and chloroplasts. J Mol Biol 266:246–268

Ogg SC, Lamond AI (2002) Cajal bodies and coilin-moving towards function. J Cell Biol 159:17–21

Omer AD, Lowe TM, Russell AG, Ebhardt H, Eddy SR, Dennis PP (2000) Homologs of small nucleolar RNAs in Archaea. Science 288:517–522

Omer AD, Ziesche S, Ebhardt H, Dennis PP (2002) In vitro reconstitution and activity of a C/D box methylation guide ribonucleoprotein complex. Proc Natl Acad Sci USA 99:5289–5294

Pelczar P, Filipowicz W (1998) The host gene for intronic U17 small nucleolar RNAs in mammals has no protein-coding potential and is a member of the 5′-terminal oligopyrimidine gene family. Mol Cell Biol 18:4509–4518

Qu LH, Henras A, Lu YJ, Zhou H, Zhou WX, Zhu YQ, Zhao J, Henry Y, Caizergues-Ferrer M, Bachellerie JP (1999) Seven novel methylation guide small nucleolar RNAs are processed from a common polycistronic transcript by Rat1p and RNase III in yeast. Mol Cell Biol 19:1144–1158

Qu LH, Meng Q, Zhou H, Chen YQ, Liang-Hu Q, Qing M, Hui Z, Yue-Qin C (2001) Identification of 10 novel snoRNA gene clusters from Arabidopsis thaliana. Nucleic Acids Res 29:1623–1630

Regalia M, Rosenblad MA, Samuelsson T (2002) Prediction of signal recognition particle RNA genes. Nucleic Acids Res 30:3368–3377

Richard P, Darzacq X, Bertrand E, Jady BE, Verheggen C, Kiss T (2003) A common sequence motif determines the Cajal body-specific localization of box H/ACA scaRNAs. EMBO J 22:4283–4293

Rivas E, Eddy SR (2000) Secondary structure alone is generally not statistically significant for the detection of noncoding RNAs. Bioinformatics 16:583–605

Rivas E, Eddy SR (2001) Noncoding RNA gene detection using comparative sequence analysis. BMC Bioinformatics 2:8

Rivas E, Klein RJ, Jones TA, Eddy SR (2001) Computational identification of noncoding RNAs in E. coli by comparative genomics. Curr Biol 11:1369–1373

Roberts TG, Sturm NR, Yee BK, Yu MC, Hartshorne T, Agabian N, Campbell DA (1998) Three small nucleolar RNAs identified from the spliced leader-associated RNA locus in kinetoplastid protozoans. Mol Cell Biol 18:4409–4417

Rozhdestvensky TS, Tang TH, Tchirkova IV, Brosius J, Bachellerie J-P, Hüttenhofer A (2003) Binding of L7Ae protein to the K-turn of archaeal snoRNAs: a shared RNA binding motif for C/D and H/ACA box snoRNAs in Archaea. Nucl. Acids. Res. 31:869–877

Russell AG, Schnare MN, Gray MW (2004) Pseudouridine-guide RNAs and other Cbf5p-associated RNAs in Euglena gracilis. RNA 10:1034–1046

Samarsky DA, Balakin AG, Fournier MJ (1995) Characterization of three new snRNAs from Saccharomyces cerevisiae: snR34, snR35 and snR36. Nucleic Acids Res 23:2548–2554

Samarsky DA, Fournier MJ (1999) A comprehensive database for the small nucleolar RNAs from Saccharomyces cerevisiae. Nucleic Acids Res 27:161–164

Samarsky DA, Fournier MJ, Singer RH, Bertrand E (1998) The snoRNA box C/D motif directs nucleolar targeting and also couples snoRNA synthesis and localization. EMBO J 17:3747–3757

Schattner P (2002) Searching for RNA genes using base-composition statistics. Nucleic Acids Res 30:2076–2082

Schattner P, Decatur WA, Davis CA, Ares M Jr, Fournier MJ, Lowe TM (2004) Genome-wide searching for pseudouridylation guide snoRNAs: analysis of the Saccharomyces cerevisiae genome. Nucleic Acids Res 32:4281–4296

Schimmang T, Tollervey D, Kern H, Frank R, Hurt EC (1989) A yeast nucleolar protein related to mammalian fibrillarin is associated with small nucleolar RNA and is essential for viability. EMBO J 8:4015–4024

Singh SK, Gurha P, Tran EJ, Maxwell ES, Gupta R (2004) Sequential 2'-O-methylation of archaeal pre-tRNATrp nucleotides is guided by the intron-encoded but trans-acting box C/D ribonucleoprotein of pre-tRNA. J Biol Chem 279:47661–47671

Smith CM, Steitz JA (1998) Classification of gas5 as a multi-small-nucleolar-RNA (snoRNA) host gene and a member of the 5'-terminal oligopyrimidine gene family reveals common features of snoRNA host genes. Mol Cell Biol 18:6897–6909

Starostina NG, Marshburn S, Johnson LS, Eddy SR, Terns RM, Terns MP (2004) Circular box C/D RNAs in Pyrococcus furiosus. Proc Natl Acad Sci USA 101:14097–14101

Steitz JA, Tycowski KT (1995) Small RNA chaperones for ribosome biogenesis. Science 270:1626–1627

Storz G (2002) An expanding universe of noncoding RNAs. Science 296:1260–1263

Szewczak LB, DeGregorio SJ, Strobel SA, Steitz JA (2002) Exclusive interaction of the 15.5 kD protein with the terminal box C/D motif of a methylation guide snoRNP. Chem Biol 9:1095–1107

Tang TH, Bachellerie JP, Rozhdestvensky T, Bortolin ML, Huber H, Drungowski M, Elge T, Brosius J, Hüttenhofer A (2002) Identification of 86 candidates for small non-messenger RNAs from the archaeon Archaeoglobus fulgidus. Proc Natl Acad Sci USA 99:7536–7541

Tang TH, Polacek N, Zywicki M, Huber H, Brugger K, Garrett R, Bachellerie JP, Hüttenhofer A (2005) Identification of novel non-coding RNAs as potential antisense regulators in the archaeon Sulfolobus solfataricus. Mol Microbiol 55:469–481

Tollervey D, Lehtonen H, Jansen R, Kern H, Hurt EC (1993) Temperature-sensitive mutations demonstrate roles for yeast fibrillarin in pre-rRNA processing, pre-rRNA methylation, and ribosome assembly. Cell 72:443–457

Tran EJ, Zhang X, Maxwell ES (2003) Efficient RNA 2′-O-methylation requires juxtaposed and symmetrically assembled archaeal box C/D and C^{prime}/D^{prime} RNPs. EMBO J 22:3930–3940

Tyc K, Steitz JA (1989) U3, U8 and U13 comprise a new class of mammalian snRNPs localized in the cell nucleolus. EMBO J 8:3113–3119

Tycowski KT, Shu MD, Steitz JA (1996a) A mammalian gene with introns instead of exons generating stable RNA products. Nature 379:464–466

Tycowski KT, Smith CM, Shu MD, Steitz JA (1996b) A small nucleolar RNA requirement for site-specific ribose methylation of rRNA in Xenopus. Proc Natl Acad Sci USA 93:14480–14485

Tycowski KT, You ZH, Graham PJ, Steitz JA (1998) Modification of U6 spliceosomal RNA is guided by other small RNAs. Mol Cell 2:629–638

Venema J, Tollervey D (1999) Ribosome synthesis in Saccharomyces cerevisiae. Annu Rev Genet 33:261–311

Verheggen C, Lafontaine DL, Samarsky D, Mouaikel J, Blanchard JM, Bordonne R, Bertrand E (2002) Mammalian and yeast U3 snoRNPs are matured in specific and related nuclear compartments. EMBO J 21:2736–2745

Vidovic I, Nottrott S, Hartmuth K, Luhrmann R, Ficner R (2000) Crystal structure of the spliceosomal 15.5kD protein bound to a U4 snRNA fragment. Mol Cell 6:1331–1342

Villa T, Ceradini F, Bozzoni I (2000) Identification of a novel element required for processing of intron-encoded box C/D small nucleolar RNAs in Saccharomyces cerevisiae. Mol Cell Biol 20:1311–1320

Vitali P, Royo H, Seitz H, Bachellerie J-P, Hüttenhofer A, Cavaillé J (2003) Identification of 13 novel human modification guide RNAs. Nucl Acids Res 31:6543–6551

Vogel J, Bartels V, Tang TH, Churakov G, Slagter-Jager JG, Hüttenhofer A, Wagner EGH (2003) RNomics in Escherichia coli detects new sRNA species and indicates parallel transcriptional output in bacteria. Nucleic Acids Res 31:6435–6443

Wang C, Meier UT (2004) Architecture and assembly of mammalian H/ACA small nucleolar and telomerase ribonucleoproteins. EMBO J 23:1857–1867

Wang C, Query CC, Meier UT (2002) Immunopurified small nucleolar ribonucleoprotein particles pseudouridylate rRNA independently of their association with phosphorylated Nopp140. Mol Cell Biol 22:8457–8466

Wang H, Boisvert D, Kim KK, Kim R, Kim SH (2000) Crystal structure of a fibrillarin homologue from Methanococcus jannaschii, a hyperthermophile, at 1.6 A resolution. EMBO J 19:317–323

Washietl S, Hofacker IL, Stadler PF (2005) Fast and reliable prediction of noncoding RNAs. Proc Natl Acad Sci USA 102:2454–2459

Wassarman KM, Repoila F, Rosenow C, Storz G, Gottesman S (2001) Identification of novel small RNAs using comparative genomics and microarrays. Genes Dev 15:1637–1651

Watanabe Y, Gray MW (2000) Evolutionary appearance of genes encoding proteins associated with box H/ACA snoRNAs: cbf5p in Euglena gracilis, an early diverging eukaryote, and candidate Gar1p and Nop10p homologs in archaebacteria. Nucleic Acids Res 28:2342–2352

Watkins NJ, Gottschalk A, Neubauer G, Kastner B, Fabrizio P, Mann M, Luhrmann R (1998) Cbf5p, a potential pseudouridine synthase, and Nhp2p, a putative RNA-binding protein, are present together with Gar1p in all H BOX/ACA-motif snoRNPs and constitute a common bipartite structure. RNA 4:1549–1568

Watkins NJ, Segault V, Charpentier B, Nottrott S, Fabrizio P, Bachi A, Wilm M, Rosbash M, Branlant C, Luhrmann R (2000) A common core RNP structure shared between the small nucleoar box C/D RNPs and the spliceosomal U4 snRNP. Cell 103:457–466

Watkins NJ, Dickmanns A, Luhrmann R (2002) Conserved stem II of the box C/D motif is essential for nucleolar localization and is required, along with the 15.5K protein, for the hierarchical assembly of the box C/D snoRNP. Mol Cell Biol 22:8342–8352

Wu P, Brockenbrough JS, Metcalfe AC, Chen S, Aris JP (1998) Nop5p is a small nucleolar ribonucleoprotein component required for pre-18 S rRNA processing in yeast. J Biol Chem 273:16453–16463

Xia L, Watkins NJ, Maxwell ES (1997) Identification of specific nucleotide sequences and structural elements required for intronic U14 snoRNA processing. RNA 3:17–26

Xu Y, Liu L, Lopez-Estrano C, Michaeli S (2001) Expression studies on clustered trypanosomatid box C/D small nucleolar RNAs. J Biol Chem 276:14289–14298

Yang C-Y, Zhou H, Luo J, Qu L-H (2005) Identification of 20 snoRNA-like RNAs from the primitive eukaryote, Giardia lamblia. Biochem Biophys Res Commun 328:1224–1231

Yang Y, Isaac C, Wang C, Dragon F, Pogacic V, Meier UT (2000) Conserved composition of mammalian box H/ACA and box C/D small nucleolar ribonucleoprotein particles and their interaction with the common factor Nopp140. Mol Biol Cell 11:567–577

Yu YT, Shu MD, Narayanan A, Terns RM, Terns MP, Steitz JA (2001) Internal modification of U2 small nuclear (sn)RNA occurs in nucleoli of Xenopus oocytes. J Cell Biol 152:1279–1288

Yuan G, Klambt C, Bachellerie JP, Brosius J, Hüttenhofer A (2003) RNomics in Drosophila melanogaster: identification of 66 candidates for novel non-messenger RNAs. Nucleic Acids Res 31:2495–2507

Zebarjadian Y, King T, Fournier MJ, Clarke L, Carbon J (1999) Point mutations in yeast CBF5 can abolish in vivo pseudouridylation of rRNA. Mol Cell Biol 19:7461–7472

Zhao X, Li ZH, Terns RM, Terns MP, Yu YT (2002) An H/ACA guide RNA directs U2 pseudouridylation at two different sites in the branchpoint recognition region in Xenopus oocytes. RNA 8:1515–1525

Zhou H, Chen YQ, Du YP, Qu LH (2002) The Schizosaccharomyces pombe mgU6–47 gene is required for 2'-O-methylation of U6 snRNA at A41. Nucleic Acids Res 30:894–902

Zhou H, Zhao J, Yu CH, Luo QJ, Chen YQ, Xiao Y, Qu LH (2004) Identification of a novel box C/D snoRNA from mouse nucleolar cDNA library. Gene 327:99–105

Nucleic Acids and Molecular Biology, Vol. 17
Wolfgang Nellen, Christian Hammann (Eds.)
Small RNAs
© Springer-Verlag Berlin Heidelberg 2005

A Computational Approach to Search for Non-Coding RNAs in Large Genomic Data

Stefan Gräf[1,4] · Jan-Hendrik Teune[1] · Dirk Strothmann[2,3] · Stefan Kurtz[2] · Gerhard Steger[1] (✉)

[1] Institut für Physikalische Biologie, Heinrich-Heine-Universität Düsseldorf, Universitätsstrasse 1, 40225 Düsseldorf, Germany
steger@biophys.uni-duesseldorf.de

[2] Abteilung für Genominformatik, Zentrum für Bioinformatik, Universität Hamburg, Bundesstrasse 43, 20146 Hamburg, Germany
kurtz@zbh.uni-hamburg.de

[3] AG Praktische Informatik, Technische Fakultät, Universität Bielefeld, 33501 Bielefeld, Germany

[4] *Present address*: EMBL Outstation, European Bioinformatics Institute, Wellcome Trust Genome Campus, Hinxton, Cambridge CB10 1SD, UK

Abstract Over the last few years several specialized software tools have been developed, each allowing a certain class of RNAs in sequence data to be found. Here we describe a general tool that allows us to specify many different non-coding RNAs and structural RNA elements by a simple pattern description language. To take into account that RNA is normally conserved in structure as well as in sequence, the pattern description language combines methods to describe sequence and structural similarities as well as further characteristics, e.g., thermodynamic constraints. Structure- and sequence-based patterns describing certain classes of RNAs are collected in a web-based pattern library. These include simple patterns, e.g., describing extrastable tetraloops and small regulatory stem-loop structures, as well as more complex patterns, for example describing pseudoknots, ribozymes, SRP RNAs, 5S RNA and selenocysteine insertion sequences. A web-based service allows a user to search the patterns from the library in sequences given by the user. Alternatively, the user can specify a pattern that is searched for in public genomic sequence data. Here we give a comprehensive introduction of the pattern language, describe how to systematically derive pattern descriptions, and show some results on purine riboswitches obtained using this computational approach.

Keywords HyPa · Motif · 5Pattern · RNA structure

1
Introduction

"Finding the hairpin in the haystack: searching for RNA motifs" is the title of an article by Dandekar and Hentze (1995); this title is still valid. The "haystack" of genomic sequences grows exponentially, the number of biologically functional RNA motifs also grows, and the search for an RNA motif is more complex than the search for protein-coding genes.

Why is it easier to find a protein-coding gene than a non-coding RNA (ncRNA) gene? A protein gene is a sequence of elements ordered along the genomic DNA: a pol II promoter, a Shine–Dalgarno sequence (in prokaryotes), an open reading frame, exons and introns (in eukaryotes), of which the boundaries have to be found, and a terminator. Up until now, such a simple structure has not been revealed for ncRNAs: an ncRNA gene may have a promoter (for any polymerase), or it may not have a promoter (e.g., if it is transcribed as part of an intron). In most cases an ncRNA has no open reading frame, and it may contain introns. Thus, each class of ncRNA has its special sequence features and, most prominently, a conserved secondary and/or tertiary structure.

Owing to the high level of diversity among ncRNAs, a number of tools were developed that allow only for searches of a single ncRNA family. A well-known example for this is tRNAscan-SE (Lowe and Eddy 1997), a program for detection of transfer RNA genes in genomic sequences. Because the development of such specialized programs anew for each family of ncRNAs requires a lot of effort, more generalized tools have been designed. These allow the user to specify the pattern to search and provide search methods for the given pattern specification. A few of these generalized pattern matching tools are listed in Table 1.

There are many aspects of generalized pattern matching to be discussed. Here we will concentrate on the process of designing a new pattern, derived from a set of "seed" RNA sequences. We will discuss refinement strategies for the designed pattern, and describe how to extend the seed RNA sequences by additional sequences that are homologous or paralogous to the seed RNA sequences. This design process usually consists of the following steps:

1. Discovery of a pattern in homologous RNA sequences. This is often done in an interactive way, involving multiple alignment programs (Frith et al. 2004, and references therein).
2. Design of the pattern description. In Sect. 3 we will describe the compilation of features common to all the RNAs from the previous step. Of course these features will allow us to discriminate the homologous RNAs from other sequences.
3. Refinement of the pattern description by searches in the seed RNA sequences set. This step is necessary to avoid false-negative hits; i.e., at least all RNAs from the seed set have to be found by genome-wide searches.
4. Search in genomic sequences.

The last step ends in three different scenarios:

(a) Only true-positive hits from the seed RNA sequences set are found. One should return to step 3 to relax the pattern description to find more hits.
(b) A large number of additional hits (many more than expected or manageable for further processing) is found. These are probably false-positive

Table 1 Tools available for pattern scanning in genomic sequences, sequence search, sequence alignment, and structure alignment

Name	URL	Reference
Pattern search		
HyPa	http://hypa.biophys.uni-duesseldorf.de/hypa/	Gräf et al. (2001)
PatScan	http://www-unix.mcs.anl.gov/compbio/PatScan/	D'Souza et al. (2003)
PatSearch	http://bighost.area.ba.cnr.it/BIG/PatSearch/	Grillo et al. (2003)
RNAmotif	http://www.scripps.edu/mb/case/casegr-sh-3.5.html	Macke et al. (2001)
Sequence search		
BLAST	http://www.ncbi.nlm.nih.gov/BLAST/Blast.cgi	Altschul et al. (1997)
FastA	http://www.ebi.ac.uk/fasta33/	Pearson (1990)
Sequence alignment		
ClustAl	http://www.ebi.ac.uk/clustalw/	Thompson et al. (1994)
	ftp://ftp.ebi.ac.uk/pub/software/unix/clustalx/	Jeanmougin et al. (1998)
PRRN	http://www.cbrc.jp/~gotoh/softdata.html	Gotoh (1996)
T-Coffee	http://igs-server.cnrs-mrs.fr/~cnotred/Projects_home_page/t_coffee_home_page.html	Notredame et al. (2000)
DiAlign	http://bibiserv.techfak.uni-bielefeld.de/dialign/	Morgenstern (1999)
Structure alignment		
AliFold	http://www.tbi.univie.ac.at/~ivo/RNA/	Hofacker (2003)
ConStruct	http://www.biophys.uni-duesseldorf.de/local/ConStruct/ConStruct.html	Lück et al. (1999)
ESSA	http://www.inra.fr/bia/T/essa/Doc/essa_home.html	Chetouani et al. (1997)
pmmulti	http://rna.tbi.univie.ac.at/cgi-bin/pmcgi.pl	Hofacker et al. (2004)
RNAforester	http://bibiserv.techfak.uni-bielefeld.de/rnaforester/	Höchsmann et al. (2003)
Slash	http://www.bioinf.au.dk/slash/	Gorodkin et al. (2001)
Information content		
slogo	http://www.cbs.dtu.dk/~gorodkin/appl/slogo.html	Gorodkin et al. (1997)

hits. One should return to step 3 to make the pattern more stringent to find fewer hits.

(c) A small number of additional hits (as expected and manageable for further processing) is found. One should try to verify these as members of the seed set by taking into account additional information which was not used in the search step. If this leads to a convincing set of hits, one may start with experimental verification.

A pattern description usually consists of different parts, each of which has different significance, i.e., different probabilities to match a random sequence. To successfully design a pattern, some basic understanding of pattern significance is very helpful. In the following section we therefore introduce the reader to some important concepts and basic results on probability of certain sequence and structural motifs. For the interested reader we refer to a more comprehensive collection of results by Staden (1989) and Karlin and Brendel (1992).

2
Probability of Patterns

For simplification, assume that characters in sequences are independently distributed with equal probability. Given this assumption, nucleic acid sequence patterns are less significant than protein patterns of identical length, because of the different sizes of the alphabets. For example, a nucleotide sequence ACGT has a probability $p_{\text{ACGT}} = p_A \cdot p_C \cdot p_G \cdot p_T = (1/4)^4 = 1/256 \approx 0.004$. That is, the sequence is expected to occur once every 256 nucleotides. The same protein sequence has a much lower probability of $p_{\text{ACGT}} = (1/20)^4 = 1/160\,000 \approx 6 \times 10^{-6}$.

Helical regions are quite significant on their own because the probability of finding a complementary base for any base is 6/16 (four Watson–Crick base pairs plus two wobble base pairs out of 16 possible base-base combinations). For example, a helix of length 5 has a probability $(3/8)^5 \approx 0.007$.

For most ncRNAs and structural RNA elements the degree of sequence conservation is lower in helical regions compared with that in loop regions. This is due to the fact that the three-dimensional conformation of a helix is maintained independently of sequence content, as long as it consists of Watson–Crick base pairs (Leontis et al. 2002; Lee and Gutell 2004). Even single wobble base pairs, mismatches, insertions, and deletions of bases are quite common in otherwise conserved helices. In contrast, loop regions are frequently involved in tertiary interactions or interactions with other macromolecules. This may have great influence on evolutionary degrees of freedom. As a consequence, helical regions are often specified by arbitrary sequences, with few mismatches, insertions, and/or deletions, whereas consensus se-

Table 2 Nucleic acid ambiguity code according to Cornish-Bowden (1985)

Code	Nucleotide	Mnemonic
A	A	Adenine
C	C	Cytosine
G	G	Guanine
U/T	U/T	Uracil
M	A, C	AMino
R	A, G	PuRin
W	A, U/T	Weak interaction
S	C, G	Strong interaction
Y	C, U/T	PYrimidine
K	G, U/T	Ketone
V	A, C, G	Not U, **V** follows U
H	A, C, U/T	Not G, **H** follows G
D	A, G, U/T	Not C, **D** follows C
B	C, G, U/T	Not A, **B** follows A
N	A, C, G, U/T	ANy

ADAM $(m = 0)$ $\Sigma = \frac{3}{128}$

ADAM $\frac{1}{4} \cdot \frac{3}{4} \cdot \frac{1}{4} \cdot \frac{1}{2} = \frac{3}{128}$

ADAM $(m = 1)$ $\Sigma = \frac{22}{128}$

BDAM $\frac{3}{4} \cdot \frac{3}{4} \cdot \frac{1}{4} \cdot \frac{1}{2} = \frac{9}{128}$

ACAM $\frac{1}{4} \cdot \frac{1}{4} \cdot \frac{1}{4} \cdot \frac{1}{2} = \frac{1}{128}$

ADBM $\frac{1}{4} \cdot \frac{3}{4} \cdot \frac{3}{4} \cdot \frac{1}{2} = \frac{9}{128}$

ADAK $\frac{1}{4} \cdot \frac{3}{4} \cdot \frac{1}{4} \cdot \frac{1}{2} = \frac{3}{128}$

ADAM $(m = 2)$ $\Sigma = \frac{52}{128}$

BCAM $\frac{3}{4} \cdot \frac{1}{4} \cdot \frac{1}{4} \cdot \frac{1}{2} = \frac{3}{128}$

BDBM $\frac{3}{4} \cdot \frac{3}{4} \cdot \frac{3}{4} \cdot \frac{1}{2} = \frac{27}{128}$

BDAK $\frac{3}{4} \cdot \frac{3}{4} \cdot \frac{1}{4} \cdot \frac{1}{2} = \frac{9}{128}$

ACBM $\frac{1}{4} \cdot \frac{1}{4} \cdot \frac{3}{4} \cdot \frac{1}{2} = \frac{3}{128}$

ACAK $\frac{1}{4} \cdot \frac{1}{4} \cdot \frac{1}{4} \cdot \frac{1}{2} = \frac{1}{128}$

ADBK $\frac{1}{4} \cdot \frac{3}{4} \cdot \frac{3}{4} \cdot \frac{1}{2} = \frac{9}{128}$

$$p_{\text{ADAM}[2,0,0]} = p_{\text{ADAM}} + p_{\text{BDAM}} + p_{\text{ACAM}} + \ldots + p_{\text{ACAK}} + p_{\text{ADBK}}$$

$$= p_{\text{ADAM}\,(m=0)} + p_{\text{ADAM}\,(m=1)} + p_{\text{ADAM}\,(m=2)}$$

$$= \frac{3}{128} + \frac{22}{128} + \frac{52}{128} = \frac{77}{128}$$

Fig. 1 Probability of finding the nucleotide pattern ADAM with up to two mismatches but without insertions or deletions, which is written as ADAM[2,0,0]. The given probabilities assume an identical and independent distribution of the four nucleotides. Classes of nucleotides are expressed by ambiguity codes (Table 2). The probability of finding ADAM without mismatch ($m = 0$) is $\frac{3}{128}$. With exactly one mismatch the probability is $\frac{22}{128}$. With exactly two mismatches the probability is $\frac{52}{128}$. The number of alternatives for the original sequences is given by the binomial coefficient $\binom{n}{m}$, with sequence length n and number of mismatches m. The probability of finding ADAM[2,0,0] is the sum of the probabilities of all alternatives. Consequently, the patterns ADAM, ADAM[1,0,0], and ADAM[2,0,0] are expected to occur about every 43, five, and two nucleotides, respectively.

quences in loops can often be described by a series of nucleotide ambiguity codes (Table 2) or by profiles (see later).

To approximate the probability of a pattern that allows for mismatches, insertions, and/or deletions, the probabilities of all alternatives have to be summed up. The number of different matches to a sequence of length n with m mismatches or deletions is given by the binomial coefficient $\binom{n}{m} = \frac{n!}{m!(n-m)!}$, or with i insertions by $\binom{n+i}{i} = \frac{(n+i)!}{i!n!}$. For a concrete example on a short sequence, see Fig. 1. The same calculations are valid for a helical region with mismatches, insertions, and deletions.

3
Design and Refinement of Patterns

We exemplify the design of a pattern for purine riboswitches (Mandal et al. 2003; Batey et al. 2004). Riboswitches are genetic regulatory elements found in the 5′ untranslated region (UTR) of bacterial messenger RNA (mRNA) that act in the absence of protein cofactors. Purine-responsive riboswitches directly bind purines and terminate transcription of genes involved in purine metabolism and transport.

The starting point of any pattern design for RNAs is a collection of homologous RNA sequences. If this collection is considered to be too small, one may try to extend it by the following approaches. Firstly, one may try to detect further homologs by pure sequence searches using, e.g., BLAST or FastA (Table 1). In most cases such a search will deliver only little additional information, since the hits will mostly be evolutionarily close homologs. A second approach is to detect additional members of the RNA group by using additional biological information like, e.g., the positional neighborhood of the RNA to special mRNAs, the location in 5′ or 3′ UTRs, and presence near the origin of plasmid replication. In the case of purine riboswitches, Mandal et al. (2003) exploited knowledge about the localization of the riboswitches in the relatively long 5′ UTR of mRNAs coding for proteins involved in purine metabolism and transport. By doing so they assembled the 31 sequences shown in Fig. 2.

When collecting more and more sequences, one has to be very careful not to add false-positive RNAs that do not belong to the group under consideration; such false-positives will probably lead to wrong conclusions in later steps of the pattern design process. Adding more true-positive sequences, however, will add support to the alignment and its derived pattern in terms of sensitivity and specificity.

Once a set of seed RNA sequences has been established, one aligns these, taking into account similarities in sequence as well as in structure. To do so, several tools are available (Table 1). Some of these produce alignments with-

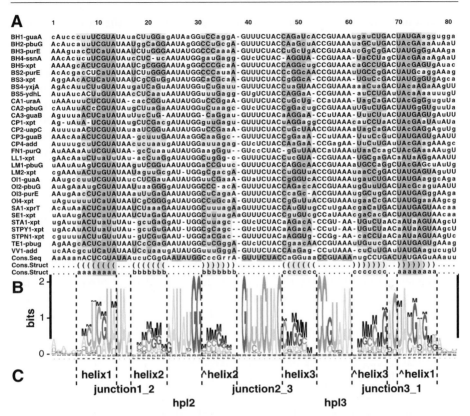

Fig. 2 Sequence and structure alignment of purine riboswitches. **A** Nomenclature (*left*) and sequences are similar to those given in Mandal et al. (2003). The structure alignment was produced with the help of ConStruct (Lück et al. 1999). Sequence characters involved in base-pairing are shown on a *gray background*; *light gray* and *dark gray* discriminate coinciding pairs from pairs covarying with the consensus, respectively. The *last three lines* show the consensus sequence and the consensus structure. In the consensus-sequence line, nucleotides are given using the ambiguity code (Table 2); the background color ranges from *white* to *dark gray*, corresponding to an increasing degree of conservation. In the lines marked *Cons.Struct*, either base pairs are given in *bracket-dot* notation (Hofacker et al. 1994) or the two regions belonging to the same helix are given by the same characters; the background color ranges from *white* to *dark gray* according to increasing probability of pairing. **B** A structure logo (Gorodkin et al. 1997) of the alignment and structure shown in **A**. The height of a character is proportional to the fraction of the observed and expected frequency. The expected frequencies ($f_A = 0.28, f_C = 0.22, f_G = 0.22, f_U = 0.28$) are for the *Bacillus subtilis* genome. When a character appears less than expected the symbol is displayed upside down. The probability of a base pair for a given alignment column is shown by the letter M in different sizes. These reflect the fractions of complementary bases at indicated base-paired positions in the alignment and the number of base pairs one would expect by chance from the distribution of nucleotides at the positions involved. **C** Nomenclature of structural elements used in the pattern description

out manual interference; others allow the user to refine the alignment step by step in an interactive way, guiding the user by a simultaneous display of sequence and structure alignment (Fig. 2).

The next step is to transfer the information derived from the sequence and structure alignment into a pattern description which conforms to the chosen pattern description language. The syntax and semantics of the chosen language is usually described with the software tool allowing us to search for the patterns (Table 1). For our example of purine riboswitches we use the language HyPaL, which is described in Tables 3 and 4. Most if not all pattern languages allow for elements such as those given in Table 3; the elements of Table 4 are special to HyPaL.

The consensus structure of our example (Figs. 2 and 3, right) consists of three helices (denoted `helix1`, `helix2`, and `helix3`); these three helices are connected in a junction (with sequence parts denoted `junction1_2`, `junction2_3`, and `junction3_1`); the hairpins (`helix1` and `helix2`) are closed by two hairpin loops (`hp12` and `hp13`). From the consensus sequence (Fig. 2A) and the statistical evaluation (Fig. 2B) it is obvious that the helical regions are less conserved in sequence than the loop regions. The only exception is the end of `helix1` (3′ part of `helix1` region and 5′ part of its complement `^helix1`) near the junction: the last four nucleotides are well conserved and, especially, the penultimate position is a U. The `helix1` has a minimum length of four base pairs (sequence BS2-*purE*). In most cases it is longer. Thus we define `helix1` as five to eight nucleotides in length ending with four defined nucleotides YRUW, and allow for one mismatch (Fig. 3, left). Helices 2 and 3 are similar in length (six to seven and five to seven nucleotides, respectively), but they contain several mismatches, insertions, and deletions. Most positions in loop regions are significantly conserved; e.g., the junction region between `helix1` and `helix2` starts with nucleotides U and A at positions 14 and 15, followed by W (i.e., A or U), which, however, is missing in one sequence (CA1-*uraA*). The remaining part is constructed in a similar way. As a final result we obtain the very specific riboswitch pattern of Fig. 3.

Designing a specific pattern is not a simple task. However, there are systematic approaches to increase pattern specificity: e.g., parts given by the nucleotide ambiguity code can be replaced by a profile; the only requirement is that there are no gaps in that part. A profile $[[s_A, s_C, s_G, s_U], ...] > t$ specifies a sequence of length at least one nucleotide via position-specific scores s for each nucleotide; a sequence matches the profile if the sum of all scores exceeds a threshold t. For example, the pattern (`junction3_1:=DY`) (Fig. 3) requires a non-C followed by a pyrimidine. That is, it matches sequences AC, AU, GC, GU, UC, and UU. Since GU and UU are not present in the corresponding alignment positions (Fig. 2), one can increase specificity by excluding them. One way of doing this is to use the profile pattern (`junction3_1:=[[27,0,3,1],[0,28,0,3]]>28`). Note the gen-

Table 3 Overview on the linguistic elements of HyPaL. A list of elements including a short example of the use of each element. For a complete definition of the language, see http://hypa.biophys.uni-duesseldorf.de/hypa/. Note that nucleotides have to be given in *uppercase*, whereas variables and functions are in *lowercase*

HyPaL	Examples for hit	Remarks
General		
Structure of a pattern		A named pattern allows for several patterns in
name=pattern;		one file; a pattern may consist of expressions
		(given below) and ends with a semicolon
Comment		All characters up to the end of the line are ignored
//		
Primary sequence		
Nucleotides	AUAAA, …	
HUMAN		All uppercase characters in IUPAC
		nomenclature
Regular expressions		
Character classes		A character class matches one character out of
[ACU]	A, C, U	the class, the caret defines a negated class
[^CU]	G, A	(match only characters not listed): H=[ACU],
		R=[AG]=[^CU]
Concatenation		
GN[GA]A GC	GGGAGC, …	
Disjunction		Specify alternatives
G|A	G, A	
Quantifiers		The brace quantifier {m, M} matches from
G{2,3}	GG, GGG	m(min) to M(max) times the previous
.{3,5}	AUG, AUGC, AUGCA, …	regular expression atom
A*G?U+	AAGU, AGU, GU, U, …	*, +, ? match zero or more, 1 or more, zero
		or one times, respectively

Table 3 continued

HyPaL	Examples for hit	Remarks
Grouping		
AA C|G AA	AAC, GAA	
AA (C|G){1,2} AA	AACAA, AACCAA, AAGAA, AAGGAA, AACGAA, AAGCAA	
(AA CC)|(GG AA)	AACC, GGAA	
AA (C|G|UU) AA	AACAA, AAGAA, AAUUAA	
Variables		
(loop:=GNRA)	GGGA, …	Variable names are in lowercase
Structure		
Reverse complement		A caret in front of a variable or element
(hx:=-{3,4}) ^hx	AUGCGCAU, …	matches its reverse complement
		Definition of special reverse complements
Base pairing		
wobble G:=[U];		
wobble U:=[G];		
wobble(GU);	UG	
Approximative elements		
(UNCG)[2,1,0]	UUCG, AUA, …	$[m,d,i]$ allows for m mismatches, d deletions,
^(AUCG)[0,1,1]	CGAU, CCGU, …	and i insertions in the expression
Profile matrices		
[[50,5,20,10], [30, 5, 60, 5]] >= 80	AA, AG, GG	Weighted profiles with $[f_A, f_C, f_G, f_U]$

Table 4 Overview of functions defined in HyPaL. For a complete definition of the language, see http://hypa.biophys.uni-duesseldorf.de/hypa/. An example for the use of the free-energy functions is the determination of the thermodynamic stability of a certain helix plus the neighboring bulge loop from a stem-loop motif:

```
motif=(stem1:=.{3,5}) (bulge:=AAA) (stem2:=.{3,5})
      (loop:=.{3,4})
      (cstem2:=^stem2) (cstem1:=^stem1)
   << freeE(stem1( bulge stem2( loop )cstem2 )cstem1)
    - freeE(stem2( loop )cstem2) < -8;
```

Function	Return value
length(v)	Length of the sequence bound to variable v
xcontent(as,v)	Relative number of occurrences of characters from variable as in the sequence bound to v
gccontent(v)	Relative number of occurrences of G and C in sequence bound to v gccontent(v)=xcontent(GC,v)
absxcontent(as,v)	Absolute number of occurrences of characters from as in the sequence bound to v
mfreeE(v)	Minimum free energy (kcal/mol) of the sequence bound to v
freeE(v)	Free energy of the RNA secondary structure specified by v in kilocalories per mole
mfeDiff(v)	freeE(v)-mfreeE(v)

```
purine_rs=
    (helix1:=.{1,4} YRUW)
    (junction1_2:=UA W?)
    (helix2:=.{6,7})
    (hpl2:=DAU(A?)NGG)
    (^helix2[3,1,0])
    (junction2_3:=A? KUNYCUA .? C)
    (helix3:=.{5,7})
    (hpl3:=CCDN N? AH)
    (^helix3[1,1,1])
    (junction3_1:=DY)
    (^helix1[1,0,0]);
```

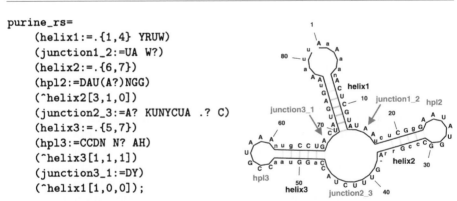

Fig. 3 Search pattern in HyPaL language (*left*) and consensus structure (*right*) for purine riboswitches based on the alignment in Fig. 2. The gray scale of lines connecting base pairs from *white* to *dark gray* corresponds to the probability of consensus pairing

eral scheme behind this method: by increasing or decreasing weights and choosing an appropriate threshold one can fine-tune the set of sequences matching at consecutive columns of the given multiple alignment. This can be done in a systematic way (Nevill-Manning et al. 1998).

Besides using the standard elements described in Table 3, one can further increase the specificity of a pattern by additional constraints. These are specified using the functions of Table 4. For example, a helix might be well conserved, but as a pattern it might be unspecific owing to mismatches, insertions, and deletions. These defects result in a thermodynamic destabilization of the helix by loop formation, but might be compensated for by an increase in the number of $G:C/G:C$ stacks. Another well-known situation is that in a larger loop the individual sequences allow for base-pairing, but differently at different positions. For both cases restrictions on the free energy ΔG_T^0 of structure formation might be helpful; i.e., sequences of low overall stability or low local stability can be excluded. Note however, that evaluating the energy functions requires computational effort proportional to m^3 for a given pattern size m.

The next step in pattern design is to verify that the pattern matches at least the sequences used for its initial design. This verification and refinement process requires several runs over the same seed data set. Since this data set is small, we expect small computing times. Note, however, that this strategy requires that the chosen pattern search program allows the user to provide his/her own data sets. This is usually easy if the program is installed locally, but is often not possible if a web-service is used. In such a case, the verification process is run over large collections of sequences (hopefully including the seed RNA sequences), resulting in long running times and in additional manual effort to distinguish true-positive from false-positive hits. Thus, a web-based approach without upload of user sequences is not appropriate.

4
Searching Patterns in Large Sequence Sets

The algorithms that are used by pattern search programs can be classified according to two different criteria. Firstly, the programs search either in raw sequence data or in a pre-computed index derived from the raw sequences. In the first case, the computational effort for each search depends on the length n of the sequence data. If a pre-computed index is used, each search requires only time depending on the pattern size m, which is smaller than n by several orders of magnitude. Of course, the pre-computation step building the index depends on n. But this has to be done only once, while there are usually many searches for different patterns. As a consequence, index-based programs are usually faster than programs searching raw sequences. Secondly, programs may or may not analyze the pattern to determine the order of the different parts of the pattern, according to their statistical significance. Such an order can be very helpful, since first searching for parts that are expected to occur rarely in the sequence data will usually improve the overall performance of the search.

A few of the points mentioned are depicted in Table 5. HyPaSearch uses a pre-computed index (Abouelhoda et al. 2004; Beckstette et al. 2004) and

Table 5 Comparison of running times for different patterns between HyPaSearch and PatScan (Table 1). Tests were performed on a computer with an AMD Opteron processor (1.8 GHz) and 7 GB RAM. The test sequence was chromosome IV of *Arabidopsis thaliana* with 1.86×10^7 nucleotides. HyPa's index creation for this sequence needs about 3-min user time. Computing times are given for 100 searches of the given pattern. The factor gives the relative computing times of PatScan to HyPaSearch. Note that PatScan finds fewer hits than HyPaSearch for certain patterns; this is based on algorithmic shortcomings of PatScan with approximative patterns

HyPaSearch		PatScan		Factor
CPU time (s)	Hits	CPU time (s)	Hits	
CUGAUGAUAA NNNNNNNN GA;				
1.2	6	235.9	6	201.6
GA NNNNNNNN CUGAUGAUAA;				
1.1	3	259.2	3	231.4
(stem:=.{20,30}) .{4} ^stem;				
651.6	20	36 923.3	20	56.7
(stem:=.{7} R) (loop:=UNCG) ^stem;				
8.3	89	375.9	89	45.3
AAAAAAAAA[1,0,0] CUGAUGAWA;				
2.3	2	394.6	2	173.8
AAAAAAAAA[1,1,1] CUGAUGAUA;				
102.7	6	1500.1	3	14.6
(stem:=G{5} W G{3}) .{4} ^stem[1,1,1];				
3.0	24	230.4	10	76.8
profile[a]				
15.1	35	1015.7	35	67.3

[a] profile = [[1,2,3,4],[1,2,3,4],[4,2,3,1],[1,4,3,1],
    [1,2,3,4],[1,2,3,4],[4,2,3,1],[1,4,3,1],
    [4,2,3,1],[1,4,3,1],[1,2,3,4],[1,2,3,4],
    [4,2,3,1],[1,4,3,1],[1,2,3,4],[1,2,3,4],
    [4,2,3,1],[1,4,3,1],[4,2,3,1],[1,4,3,1]]>=72;

search-order optimization. The total search time for the example patterns of Table 5 sums up to only 13 min. In contrast, PatScan searches directly in the raw sequence, but needs more than 11 h.

It is a good strategy to start the pattern search in this phase on a fully sequenced and annotated chromosome or genome that contains at least one of the sequences from the seed RNA sequence set. This restricted search has several advantages. Firstly, one knows that the pattern has to be found at least once. Secondly, verification of any additional hit sequence(s) is relatively easy owing to the available annotation. Finally, this search should allow for a reasonable estimation of the running time for the corresponding search in the full EMBL database or GenBank.

A final search in the full EMBL database or GenBank is quite laborious and cumbersome in most cases. Owing to the fact that these databases are redundant, one usually finds many redundant hits. If a hit occurs more than once in regions that are not annotated, it might be helpful to additionally extract about 100 additional nucleotides at both ends of the hit, either by special command options of the search program or by corresponding unspecific nucleotides added to the pattern. An alignment of such extended hits will allow for identification of redundant sequences.

A search with the purine riboswitch-specific pattern `purine_rs` (Fig. 3) in the bacterial subdivision of the EMBL database (version 78; Kulikova et al. 2004) finds 253 hits. As expected, all sequences from the seed RNA sequences set (Fig. 2) are found. There are also many redundant hits, and additional hits. All additional hits are true-positives according to additional criteria. That is, they are located in 5′ UTRs of mRNAs involved in purine metabolism and transport. One result is remarkable: there are one to five hits in certain full bacterial genomes. This may be interpreted as follows: either the bacteria make use of the purine riboswitches to different extents, or those bacteria with a lower number of hits possess purine riboswitches that do not match the pattern `purine_rs`. To verify these hypotheses, one will have to search for protein homologs of the already known proteins and analyze their 5′ UTRs. This may also give hints on how to relax the pattern description, if necessary.

5
WWW Tools

The pattern search programs of Table 1 may be used as standalone programs with their command-line interface. Many users prefer to use a program via a graphical user interface and/or a WWW interface. To support the stepwise process as explained in the two previous sections, the following features should be supported:

1. The interface guides the user to simplify input of a pattern which conforms with the syntax of the pattern language.
2. A pattern can be visualized by a graph, to allow the user to check the pattern for syntactical correctness and agreement with the intended pattern.
3. A restricted upload of sequences and searches within these is possible in order to verify the pattern.
4. Results of a search can be selected and sorted according to several criteria.
5. To simplify the interpretation of results, description and features of hit sequences are available via a hot link to the appropriate sequence database.
6. For further processing of hit sequences, direct links to standard tools like RNA structure prediction (e.g., `RNAfold`; Hofacker 2003) or sequence alignment (e.g., `ClustAl`; Jeanmougin et al. 1998) are available.

These features are available in the HyPa web portal (Fig. 4).

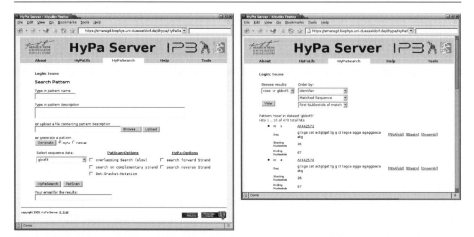

Fig. 4 Screenshots of the user interface of HyPaSearch (http://hypa.biophys.uni-duessel dorf.de/hypa/HyPaSearch). *Left* The user is able to type in or to upload a pattern in HyPaL or PatScan syntax, to select different program options, and then to search in available or user-provided sequence data. *Right* Results are shown in user-defined order with different options for further processing

Fig. 5 Screenshot of an entry in HyPaLib. For a description of the HyPaLib items see Gräf et al. (2001)

Based on HyPaL and the HyPaSearch webserver, we are developing Hy-PaLib, a "hybrid pattern library", that already contains about 60 sequential and structural elements characteristic for different classes of RNA. HyPaLib is available as a plain text file, formatted according to the general rules of the EMBL and related databases. We also provide a version in HTML format with hyperlinks to sequence and citation databases (Fig. 5). Similar to the patterns from the Prosite database (Hulo et al. 2004), patterns from Hy-PaLib should help to reliably identify ncRNAs and structural RNA elements in sequences.

We believe that the annotation of ncRNAs requires approaches that systematically reveal common features of large sets of sequences, visible in similarities in the sequence and in the secondary structure. In this review we have described such an approach. We are now in a situation where the appropriate software tools are becoming available to support this approach. We hope that this motivates other researchers to follow or improve our approach by analyzing their favorite set of RNA sequences. In many cases, a pattern may be derived which allows us to reveal new secrets in the world of RNA.

Acknowledgements We thank Michael Schmitz for stimulating discussions about the probability calculations. This work was supported by a grant of the Deutsche Forschungsgemeinschaft to S.K. and G.S.

References

Abouelhoda MI, Kurtz S, Ohlebusch E (2004) Replacing suffix trees with enhanced suffix arrays. J. Discrete Algorithms 2:53–86

Altschul SF, Madden TL, Schaffer AA, Zhang J, Zhang Z, Miller W, Lipman DJ (1997) Gapped BLAST and PSI-BLAST: a new generation of protein database search programs. Nucleic Acids Res 25:3389–3402

Batey RT, Gilbert SD, Montange RK (2004) Structure of a natural guanine-responsive riboswitch complexed with the metabolite hypoxanthine. Nature 432:411–415

Beckstette M, Strothmann D, Homann R, Giegerich R, Kurtz S (2004) PoSSuMsearch: fast and sensitive matching of position specific scoring matrices using enhanced suffix arrays. In: Giegerich R, Stoye J (eds) Proceedings of the German conference on bioinformatics (GCB 2004), vol 53 GI, Bielefeld, Germany, 4–6 October 2004, pp 53–64

Chetouani F, Monestié P, Thébault P, Gaspin C, Michot B (1997) ESSA: an integrated and interactive computer tool for analysing RNA secondary structure. Nucleic Acids Res 25:3514–3522

Cornish-Bowden A (1985) Nomenclature for incompletely specified bases in nucleic acid sequences: recommendations 1984. Nucleic Acids Res 13:3021–3030

Dandekar T, Hentze MW (1995) Finding the hairpin in the haystack: searching for RNA motifs. Trends Genet 11:45–50

D'Souza M, Larsen N, Overbeek R (1997) Searching for patterns in genomic data. Trends Genet 13:497–498

Frith MC, Hansen U, Spouge JL, Weng Z (2004) Finding functional sequence elements by multiple local alignment. Nucleic Acids Res 32:189–200

Gorodkin J, Heyer LJ, Brunak S, Stormo GD (1997) Displaying the information contents of structural RNA alignments: the structure logos. Comp Appl Biosci 13:583–586

Gorodkin J, Stricklin SL, Stormo GD (2001) Discovering common stem-loop motifs in unaligned RNA sequences. Nucleic Acids Res 29:2135–2144

Gotoh O (1996) Significant improvement in accuracy of multiple protein sequence alignments by iterative refinement as assessed by reference to structural alignments. J Mol Biol 264:823–838

Gräf S, Strothmann D, Kurtz S, Steger G (2001) HyPaLib: a database of RNAs and RNA structural elements defined by hybrid patterns. Nucleic Acids Res 29:196–198

Grillo G, Licciulli F, Liuni S, Sbisa E, Pesole G (2003) PatSearch: a program for the detection of patterns and structural motifs in nucleotide sequences. Nucleic Acids Res 31:3608–3612

Höchsmann M, Töller T, Giegerich R, Kurtz S (2003) Local similarity in RNA secondary structures. In: 2nd IEEE Computer Society Bioinformatics Conference (CSB 2003), 11–14 August 2003, Stanford, CA, USA IEEE Computer Society, p 159–168

Hofacker IL (2003) Vienna RNA secondary structure server. Nucleic Acids Res 31:3429–3431

Hofacker IL, Fontana W, Stadler PF, Bonhoeffer S, Tacker M, Schuster P (1994) Fast folding and comparsion of RNA structures. Monatsh Chem 125:167–188

Hofacker IL, Bernhart SHF, Stadler PF (2004) Alignment of RNA base pairing probability matrices. Bioinformatics 20:2222–2227

Hulo N, Sigrist CJA, Le Saux V, Langendijk-Genevaux PS, Bordoli L, Gattiker A, De Castro E, Bucher P, Bairoch A (2004) Recent improvements to the PROSITE database. Nucleic Acids Res 32:D134–D137

Jeanmougin F, Thompson JD, Gouy M, Higgins DG, Gibson TJ (1998) Multiple sequence alignment with Clustal X. Trends Biol Sci 23:403–405

Karlin S, Brendel V (1992) Chance and statistical significance in protein and DNA analysis. Science 257:39–49

Kulikova T, Aldebert P, Althorpe N, Baker W, Bates K, Browne P, van den Broek A, Cochrane G, Duggan K, Eberhardt R, Faruque N, Garcia-Pastor M, Harte N, Kanz C, Leinonen R, Lin Q, Lombard V, Lopez R, Mancuso R, McHale M, Nardone F, Silventoinen V, Stoehr P, Stoesser G, Tuli MA, Tzouvara K, Vaughan R, Wu D, Zhu W, Apweiler R (2004) The EMBL nucleotide sequence database. Nucleic Acids Res 32:27–30

Lee JC, Gutell RR (2004) Diversity of base-pair conformations and their occurrence in rRNA structure and RNA structural motifs. J Mol Biol 344:1225–1249

Leontis NB, Stombaugh J, Westhof E (2002) The non-Watson–Crick base pairs and their associated isostericity matrices. Nucleic Acids Res 30:3497–3531

Lowe TM, Eddy SR (1997) tRNAscan-SE: a program for improved detection of transfer RNA genes in genomic sequence. Nucleic Acids Res 25:955–964

Lück R, Gräf S, Steger G (1999) ConStruct: a tool for thermodynamic controlled prediction of conserved secondary structure. Nucleic Acids Res 27:4208–4217

Macke TJ, Ecker DJ, Gutell RR, Gautheret D, Case DA, Sampath R (2001) RNAMotif, an RNA secondary structure definition and search algorithm. Nucleic Acids Res 29:4724–4735

Mandal M, Boese B, Barrick JE, Winkler WC, Breaker RR (2003) Riboswitches control fundamental biochemical pathways in Bacillus subtilis and other bacteria. Cell 113:577–586

Morgenstern B (1999) DIALIGN 2: improvement of the segment-to-segment approach to multiple sequence alignment. Bioinformatics 15:211–218

Nevill-Manning CG, Wu TD, Brutlag DL (1998) Highly specific protein sequence motifs for genome analysis. Proc Nat Acad Sci USA 95:5865–5871

Notredame C, Higgins DG, Heringa J (2000) T-Coffee: a novel method for fast and accurate multiple sequence alignment. J Mol Biol 302:205–217

Pearson WR (1990) Rapid and sensitive sequence comparison with FASTP and FASTA. Methods Enzymol 183:63–98

Staden R (1989) Methods for calculating the probabilities of finding patterns in sequences. Comp Appl Biosci 5:89–96

Thompson JD, Higgins DG, Gibson TJ (1994) CLUSTAL W: improving the sensitivity of progressive multiple sequence alignment through sequence weighting, position-specific gap penalties and weight matrix choice. Nucleic Acids Res 22:4673–4680

Nucleic Acids and Molecular Biology, Vol. 17
Wolfgang Nellen, Christian Hammann (Eds.)
Small RNAs
© Springer-Verlag Berlin Heidelberg 2005

Experimental Strategies for the Identification and Validation of Target RNAs that Are Regulated by miRNAs

Alexandra Boutla (✉) · Martin Tabler

Institute of Molecular Biology and Biotechnology, Foundation for Research
and Technology—Hellas, P.O. Box 1527, 71110 Heraklion/Crete, Greece
boutla@embl.de

Abstract Micro-RNAs (miRNAs) are single-stranded RNA molecules of about 20 nucleotides and represent a class of small non-coding regulatory RNAs found in higher eukaryotes across kingdoms. miRNAs originate from endogenous chromosomal genes that encode a transcript, called pre-miRNA, of about 70 nucleotides in length. Several pre-miRNAs can be combined in a polycistronic pri-miRNA. Both precursors are processed by specific ribonucleases: Drosha cleaves a pri-miRNA to several pre-miRNAs, which are processed by a Dicer nuclease into the mature miRNA. The miRNAs themselves interfere with gene expression by base-pairing with a messenger RNA (mRNA) target, but one miRNA can bind in a specific manner to several mRNAs; therefore, miRNAs form a regulatory network that controls gene expression. In this way they are believed to determine tissue-specific gene expression and to act as checkpoints for developmentally important processes. Only a few plant miRNAs show a perfect match to their mRNA target, but the majority of miRNAs, including mammalian miRNAs, regulate translation by imperfect base-pairing. Each species is expected to have a couple of hundred miRNAs and only a few validated targets are known so far. Several attempts have been made to identify miRNA targets by bioinformatics. However, the flexibility of base-pairing is high and real target mRNAs are hard to predict. As an alternative, we have developed experimental strategies to identify mRNA targets, which will be summarised in this chapter. Two approaches make use of the antisense sequence of a miRNA in the form of a DNA oligonucleotide to either inactivate a specific miRNA or to direct a PCR in combination with a second DNA oligonucleotide that matches to the vector sequence. This reaction can be used in a complementary DNA (cDNA) library, preferentially a normalised cDNA library or on cDNAs generated directly from mRNAs. The amplified product is cloned and sequenced, so that the gene can be identified from the data base. In this way we could identify several promising target genes in *Drosophila melanogaster* that are regulated by either miR-2 or miR-13. As an ultimate proof, target validation is required. For that purpose the 3′ untranslated region of a potential miRNA-regulated gene is fused to the coding region of a reporter gene. The binding site of the miRNA is mutagenised and the expression of the reporter monitored. The combination of these strategies allows us to identify miRNA-regulated mRNAs.

1
Introduction

Genes are defined as heredity units and can be divided into two major groups: genes that code for a protein and "non-coding" genes that are involved in regulatory processes or encode a functional RNA. Typical non-coding RNA genes in eukaryotes are ribosomal, transfer, splicosomal and small nuclear RNAs, which are involved in either RNA processing or other basic processes facilitating directly or indirectly gene expression. Recently, micro-RNAs (miRNA) were discovered as a further group of non-coding RNAs, and have regulatory function by modulating gene expression at the translational level.

1.1
History

In the late 1980s, Ambros and coworkers discovered several heterochronic genes that affected the timing of the larva stages in *Caenorhabditis elegans* (Abbott 2003; Ambros 1989; Ambros and Horvitz 1984; Ambros and Moss 1994). Mutations in the heterochronic gene *lin-4* affected the timing of the transition of larva stage L1 to L2 during the nematode development and another heterochronic gene *let-7* affected the later stages of the larva development. According to the original findings, *lin-4* and *let-7* genes do not code for a protein, but for two small RNAs of 22 nucleotides that are excised from a longer precursor transcript of about 60 nucleotides. Because of their small size and their specific temporal regulation they were denominated *small temporal RNAs* (stRNAs) (Lee et al. 1993; Olsen and Ambros 1999).

At that time stRNAs appeared to be a rather exotic exception of regulatory small RNAs specific for *C. elegans*. However, almost a decade later perfect homologues of *let-7* were discovered in flies, zebrafish and humans, and in some of these organisms stage-specific expression of *let-7* was observed (Pasquinelli et al. 2000). Thereafter, three laboratories discovered independently in flies and nematodes some dozen genome-derived single-stranded small RNA sequences resembling the *lin-4* and *let-7* of *C. elegans* (Lagos-Quintana et al. 2001; Lau et al. 2001; Lee and Ambros 2001). These RNAs were discovered unexpectedly in an effort to clone and characterise small interfering RNA (siRNA). This new class of small RNAs was termed miRNA and in retrospect it was clear that stRNAs, like *lin-4* and *let-7*, were part of this RNA class.

1.2
Biogenesis

Today several hundreds miRNA genes have been discovered in worms, flies, plants and mammals (Ambros et al. 2003; Lagos-Quintana et al. 2002, 2003;

Lim et al. 2003; Mallory and Vaucheret 2004; Reinhart et al. 2002; Wang et al. 2004). Some miRNAs are conserved across species. Therefore a unique nomenclature system and criteria for classification have been proposed (Ambros et al. 2003).

Lately it was found that most, if not all, miRNA genes are transcribed by DNA-dependent RNA polymerase II (polII), resulting in a primary transcript (pri-miRNA) that is capped and polyadenylated (Lee et al. 2004). The primary transcript is then subject to a two-step processing reaction to generate the mature form of a miRNA (Lee et al. 2002). This reaction takes place in the nucleus (possibly with the exception of plants, where the precursors are believed to be first exported; Papp et al. 2003). The first reaction is the cleavage of the pri-miRNA to pre-miRNA by Drosha, a nuclear RNaseIII-like protein, (Lee et al. 2003). The pre-miRNA has a characteristic hairpin-like secondary structure; it is approximately 70 nucleotides in length and contains one specific miRNA sequence, either on the 5′ or on the 3′ region of the hairpin. The miRNA is excised from the pre-miRNA in the cytoplasm by a second specific RNaseIII-like nuclease, Dicer (Lund et al. 2004; Park et al. 2002; Xie et al. 2004). Some organisms have miRNA-specific Dicer enzymes (*Drosophila melanogaster*, *Arabidopsis thaliana*), while mammals use the same Dicer that is involved in processing double-stranded RNA to siRNA during RNA interference (RNAi) (Lee et al. 2004; Schauer et al. 2002).

miRNAs control gene expression by binding to complementary sites on target messenger RNAs (mRNAs) and negatively regulate translation efficiency or the stability of the target. The degree of complementarity between miRNA and its target mRNA determines the mode of inhibition. Perfect or almost perfect pairing induces cleavage of the target RNA via the RNAi pathway, whereas imperfect pairing in the central part of the duplex is believed to arrest translation (Doench et al. 2003; Hutvagner and Zamore 2002).

A combination of computational and experimental techniques identified a series of potential miRNA:mRNA pairs, suggesting that miRNA participates in many regulatory cellular pathways (Enright et al. 2003; Jones-Rhoades and Bartel 2004; Kiriakidou et al. 2004; Lewis et al. 2003; Rajewsky and Socci 2004; Stark et al. 2003). The analysis of miRNA targets revealed differences in plants and animals. In plants miRNAs could be identified that bind to their targets with perfect complementarity or only a few mismatches, and that mediate cleavage of the mRNA. Moreover, plant miRNAs have target sites within the coding sequence. However, this does not mean that all plant miRNAs will have almost perfect target pairing and that all act by RNA cleavage. On the other hand, in animals, the binding sites of the miRNA are usually located in the 3′ untranslated region (UTR) of their mRNA target and they bind with only partial complementarity, so that they control gene expression by blocking translation, most likely in a concentration-dependent and reversible manner.

1.3
Computer Prediction

Several computer algorithms have been designed with the purpose of screening sequenced genomes in order to identify new miRNA genes. These algorithms have three basic steps: first, they screen genomic sequences for hairpin-like structures residing in intergenic sequences and introns; second, sequence filters are applied for the refinement of the results, based on rules that are derived from the known miRNA genes; third, the potential miRNAs identified are analysed by sequence conservation filters of whether they can be found in closely and distantly related species (Grad et al. 2003; Jones-Rhoades and Bartel 2004; Lai et al. 2003; Lim et al. 2003). Information about individual miRNAs from various organisms, including sequence data of precursors and mature forms, as well as genomic positions is accessible in a database from the Sanger Institute, "The miRNA registry" http://www.sanger.ac.uk/Software/Rfam/mirna/index.shtml (Griffiths-Jones 2004).

On the basis of bioinformatics analysis, together with experimental validation of these results, several hundred miRNA genes have been characterised (Ambros et al. 2003). Further analysis showed that miRNA genes are abundant in the genomes and in every case the primary transcripts form hairpins which are processed to produce the mature miRNAs. There is no specific rule about the position of miRNA genes in the genome. They are located mainly in intergenic regions, but some miRNAs are located in intron sequences. MiRNA genes can be found both in isolated loci and also in clustered structures resembling the operons of prokaryotic genes.

1.4
Families of miRNAs

miRNAs can be categorised in families according to their sequence and their target RNAs. Each miRNA family may contain several members that differ in a few nucleotides, sometimes only in one or two nucleotides. Related miRNA may form operon-like structures and are processed from polycistronic pri-miRNAs, but there are also some family members that are encoded on different chromosomes (Aravin et al. 2003). It seems that there is redundancy in miRNA sequences located at different genomic positions. Despite this redundancy, a specific locus may be connected with a specific phenotype. For example, the human *miR-15/16* cluster, appears in the genome twice on different chromosomes, but only deletion of the 13q14 locus is connected with B cell chronic leukemia (Calin et al. 2002, 2004a, 2004b). Similarly, the *miR-165/166* cluster in plants appears several times in the genome, but each locus has a specific expression pattern, presumably owing to different promoters (Juarez et al. 2004). Clustered miRNAs that are processed from a polycistronic transcript usually share common expression patterns like *miR-124, miR-100* and *let-7* in *Drosophila* (Aravin et al. 2003).

1.5
Target Prediction by Data Base Search

Several computer programs have been described that predict target mRNAs for specific miRNAs (Enright et al. 2003; John et al. 2004; Kiriakidou et al. 2004; Lewis et al. 2003; Rhoades et al. 2002; Stark et al. 2003). These programs identified a large number of possible miRNA:mRNA pairs. However, so far only five animal miRNA targets have been verified experimentally (Boutla et al. 2003; Brennecke et al. 2003; Lee et al. 1993; Olsen and Ambros 1999; Slack et al. 2000; Xu et al. 2003).

1.6
Genetic Identification

The stRNAs were identified genetically, since a defect in *lin-4* caused a heterochronic mutation (Lee et al. 1993). However, genetic defects in miRNA genes appear rare, not only because of the small size of the genes, but also because of the redundancy of many miRNA genes. In some cases a single miRNA locus may be connected with a specific phenotype; in other cases many different loci contribute to one developmental phenotype (Juarez et al. 2004). Thus, the genetic analysis for a specific miRNA gene may be difficult, since many areas in the genome have to be deleted simultaneously and directed genomic deletions or mutagenesis are not available in many organisms.

1.7
Experimental Strategies for Target Identification

Computer prediction programs seek for strong interactions between miRNA and target mRNA to identify matching RNA–RNA pairs. This does not necessarily reflect the cellular situation. For example, the genetically verified interaction of *lin4–lin14* is relatively weak. Computer algorithms that will identify *lin4–lin14* types of interactions would be likely to predict a large number of false-positive interactions. Thus, there is the need for experimental techniques for target identification rather than relying solely on computer predictions. In the following, we describe three experimental approaches. In the first strategy, miRNA genes are inactivated to obtain indications of what target genes they may regulate. This will test for phenotypic effects caused by the upregulation of the target genes. The second approach tries to identify miRNA/mRNA interactions via actual, rather than predicted base interactions, and the third method outlines how to validate targets regardless of whether they are identified experimentally or by computer prediction.

2
Post-Transcriptional Inhibition of miRNA Function by Antisense Sequences

This strategy takes advantage of the imperfect binding of miRNA to its target mRNA. An artificially introduced nucleic acid with perfect complementarity will be a better target and will effectively out-titrate the miRNA. If the antisense sequence, which could be either DNA or RNA, is offered in high concentration, all the authentic miRNA binding sites in their corresponding mRNA targets should stay free, resulting in translational upregulation of the target gene (Fig. 1a). The resulting phenotypic effect may allow for conclusions about the target genes involved.

Using this strategy, we intended to inactivate miRNA function by injecting into *Drosophila* embryos DNA oligonucleotides of antisense polarity (anti-miDNA) for 11 known miRNAs (miR-1, miR-2b, miR-3, miR-4, miR-5, miR-6, miR-7, miR-8, miR-9, miR-11 and miR-13a) (Boutla et al. 2003), which are expressed in the early stages of embryo development (Lagos-Quintana et al. 2001). We expected that the high concentration of the DNA oligonucleotide would be sufficient to out-titrate the mature form of the miRNAs, preventing them from hybridising to their RNA target.

Four out of the 11 anti-miDNA assayed caused developmental phenotypes in the *Drosophila* embryos and seven showed no visually detectable defects. However, about 25% of the embryos injected with the anti-miDNA-1 and anti-miDNA-3 showed mild developmental defects, although without a clearly defined phenotype. Further, in about two thirds of the injected embryos with the anti-miDNA-2b and anti-miDNA-13a we observed for both miRNAs the same specific developmental defect, which can be described as cuticle holes at the posterior end of the embryo that do not affect the denticle formation and the overall body pattern. In most of the cases where this phenotype was observed, it was connected to head deformations (Boutla et al. 2003).

In this context it was of interest that earlier studies in *Drosophila* had identified several short sequence motifs in the 3′ UTR of some mRNAs that are involved in translational regulation, for example the *K*-box, the *GY*-box and the *Brd*-box (Lai et al. 1998; Lai and Posakony 1997). After the discovery of miRNAs in *Drosophila,* Lai (2002) observed that some of them showed perfect or almost perfect sequence complementarity with their six to eight 5′-terminal nucleotides to those translation regulation motifs. This included miR-2, miR-6, miR-11 and miR-13, which we had tried to inactivate by antisense DNA and which have complementarity via the terminal eight bases to the *K*-box. This prompted us to term this section of the miRNA the "family" motif (Boutla et al. 2003). Further observations suggested that an almost perfect match of the 5′-terminal domain of miRNAs to the target mRNA is essential for its regulatory function (Doench and Sharp 2004). Target prediction algorithms now take this property into account as well.

miR-2 and miR-13 do not only belong to the *K*-box family, but show short regions of sequence identity between each other. Moreover, several sequence variants of miR-2 and miR-13 can be found in *Drosophila melanogaster* (three forms of miR-2 and two of miR-13), they are encoded in different loci, differ by only one nucleotide, (among the members of the same family) (Aravin et al. 2003; Griffiths-Jones 2004) and are expressed throughout the embryonic development of the fly. The *K*-box target motif is found in several developmentally important *Drosophila* genes, like the *enhancer of split* complex (Lai et al. 1998), so that developmental defects by inactivation of miR-2 and miR-13 are in agreement with the biological role of the target genes. Although we tested with the antisense inactivation method further members of the *K*-box miRNA family, like miR-6 and miR-11, we did not observe any specific developmental phenotypes. One explanation may be that other closely related miRNAs that occur in the genome can complement the function, which may also be the reason for their occurrence. In order to test for this possibility, we performed injections with mixed anti-miDNA that would out-titrate concurrently some or all of the members of the *K*-box miRNA family. However, only in those cases where anti-miDNA for miR-2 and/or miR-13 were included, could we observe a phenotype which was identical to that observed when miR-2 or miR-13 was injected alone. We confirmed the specificity of the antisense inactivation by repeating the injection experiment with an anti-miDNA-13 carrying two mismatches when paired with the miR-13 RNA. This DNA oligonucleotide was no longer able to induce the phenotypic defect. The fact that only some of the DNA oligonucleotides tested resulted in specific phenotypes in the embryo of *Drosophila* either reflects a redundancy in the function of a miRNA or indicates that certain miRNAs are only required in a specific time window of the developmental process. It is also conceivable that some miRNAs are used for a "fine-tuning" of gene expression.

Although our antisense approach was successful with conventional DNA oligonucleotides complementary to miRNA, this technique can be improved by using antisense oligonucleotides with chemical modifications. This approach has been reported for *Drosophila*, *C. elegans* and HeLa cells.

Zamore and his group (Hutvagner et al. 2004) tried to downregulate the *lin-4* and *let-7* function by injecting 2'-*O*-methyl antisense oligonucleotides complementary for *lin-4* and *let-7* into the gonads of hermaphrodite nematodes. The progeny of the injected nematodes did not show any developmental defects that were connected to *lin-4* or *let-7* downregulation. The authors supposed that the 2'-*O*-methyl oligonucleotides were not efficiently transmitted to the progeny of the injected animals. To bypass this problem they tested for phenotypes in the injected nematode larvae rather than in the progeny. However, injection into animals of larval stages L1/L2 was lethal. When Hutvagner et al. (2004) injected the chemically modified antisense oligonucleotide at later larval stages (L2/L3), the maturest larvae survived the procedure and were assayed for the *let-7* inhibition function. All the phe-

A   Out-titration of miRNA by antisense DNA

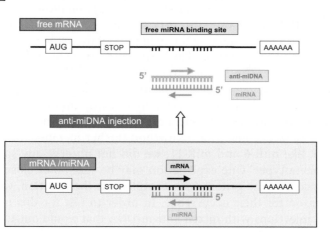

B   PCR strategy for target identification

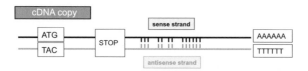

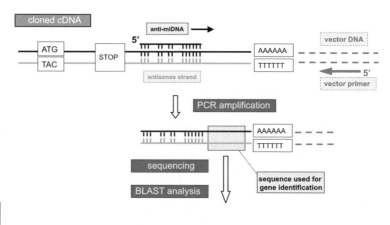

C

Target validation

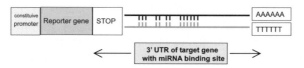

◀ **Fig. 1** Summary of the experimental strategies described in this article. The box under **a** shows the micro-RNA (*miRNA*) paired to the 3′ untranslated region (*UTR*) within a target messenger RNA (*mRNA*). Note the continuous base-pairing at the 5′ end of the miRNA. This region corresponds to the family motif of the miRNA, where only few mismatches occur. The *top* of **a** shows the effect of providing an anti-miDNA oligonucleotide: the interaction with the artificial substrate is stronger owing to perfect base-pairing, so that the miRNA binding site within the target mRNA remains free. The PCR strategy is outlined in **b**. The complementary DNA (*cDNA*) copy of the target mRNA is given with the sense and antisense strands detailed. If a cloned cDNA is available (*below*), a DNA oligonucleotide anti-miDNA with antisense polarity to the miRNA (i.e. the same polarity as the mRNA) can initiate via its 3′ terminus (that is complementary to the family motif) a PCR, together with a primer that is specific for vector sequences. The amplified PCR product is outlined schematically *below*. Because the anti-miDNA oligonucleotide will be incorporated into the cDNA there is no direct information on the nature of the miRNA binding site. Of particular interest is the sequence information of the part downstream of the miRNA binding site (*shaded area*). These data can be used for basic local alignment search tool (*BLAST*) analysis to identify the nature of the cloned gene. Once the gene has been identified, the authentic miRNA-binding site can be located in the published sequence. **c** summarises the strategy of target validation. The 3′ UTR of the identified target gene is fused to the coding region of a reporter gene, which itself is expressed from a constitutive promoter. The binding site for the miRNA can be mutated and the effect on reporter gene expression monitored

notypes associated with injection of the *let-7* complementary 2′-*O*-methyl oligonucleotide were consistent with the loss of *let-7* activity (Reinhart et al. 2000). *let-7* represses translation of *lin-41* mRNA by binding to a partially complementary site in the *lin-41* 3′ UTR (Reinhart et al. 2000; Slack et al. 2000; Vella et al. 2004). Many of the phenotypes observed after the injection of the *let-7* complementary oligonucleotides resembled the phenotypes caused by LIN-41 protein overexpression. This actually proved that the inhibition of *let-7* caused increased translation of the *lin-41* mRNA, which produced the relevant phenotypes.

Similar experiments have been performed for inactivation of miRNA function in HeLa cells. 2′-*O*-Methyl oligonucleotide complementary to the miR-21 caused specific inhibition of its function both in HeLa cell extracts and in cultures as was reported from similar experiments for *lin-4* and *let-7* (Hutvagner et al. 2004; Meister et al. 2004).

Antisense inactivation of miRNA is a powerful tool to test for essential functions, because it generates a phenotypic mutant. However, the interpretation of the results may be difficult if corresponding mutant phenotypes have not been previously reported. The downregulation of a miRNA will actually cause the upregulation of its target mRNA, so that the resulting phenotype mimics an overexpression of the target mRNA. For example, miRNAs are known to regulate pro-apoptotic genes like the miRNA *bantam* and its target *hid* in *Drosophila* (Brennecke et al. 2003). In such cases out-titration of *bantam* will lead to increased apoptosis, because of the de-repression of *hid*.

3
Direct Identification of Target mRNA by a PCR Strategy
Mimicking the miRNA Interaction

As already outlined, the 5′ end of the miRNAs often tends to bind with perfect complementarity to the mRNA target. The importance of this region was recognised by Lai (2004) and was confirmed by Doench and Sharp (2004). This is the region that we called the family motif. We exploited this mode of interaction between miRNA and target mRNA to initiate a PCR via the family motif (Boutla et al. 2003).

Owing to its location at the 5′ end of the miRNA, a DNA oligonucleotide of the same polarity as the miRNA cannot initiate a reverse transcription (RT) or a PCR. However, instead of mRNA, a cloned double-stranded DNA copy of the target RNA can be used for PCR amplification. In this case, a DNA oligonucleotide of antisense polarity to miRNA can be used (Fig. 1b). Now, the sequences corresponding to the family motif are at the 3′ end of this DNA antisense oligonucleotide, so that the six to eight 3′-terminal nucleotides fit well to the antisense DNA strand of the target mRNA. This tight interaction with the template DNA at the 3′ end of the primer will allow the initiation of a PCR. Furthermore, we selected the experimental conditions for the PCR so that an accidental match of six to eight nucleotides at the 3′ end alone would not be sufficient to initiate the PCR. The 5′-terminal nucleotides of the anti-miDNA had to contribute as well to a specific interaction, but here mismatches were allowed. We performed the PCR on a complementary DNA (cDNA) library cloned in a bacteriophage vector originating from early *Drosophila* embryos. As a reverse primer we used a DNA oligonucleotide that binds to vector sequences. Owing to the reduced specificity of the anti-miDNA oligonucleotide at the initial phase of the reaction, we performed a PCR with two annealing temperatures: initially about 10–15 °C below the T_m value for the first four or five cycles, followed by annealing temperatures close to the T_m value for the residual amplification cycles (Boutla et al. 2003). The PCR products were cloned and sequenced and the corresponding genes were identified in data bases by basic local alignment search tool (BLAST) analysis (Altschul et al. 1997).

In all cloned cDNAs we could identify the binding site of the miRNA and in almost all of them the 3′ terminus of the anti-miDNA primer had almost perfect complementarity. We performed this analysis for *Drosophila* miRNAs 1, 3 and 13a on two different *Drosophila* cDNA libraries. From this analysis, several potential miRNA targets were identified. The miRNA binding sites were mostly located in the 3′ UTR of a target gene, but the majority of the genes were of unknown function. We also identified some sequences that were annotated as introns or intergenic areas; this was surprising as these sequences were found in a cDNA library. It is possible that they represent further miRNA genes that belong to the K-box family.

Although no relevant experimental data are available yet, the PCR-screening method could be likewise used to identify targets from plant cDNA libraries. Plant miRNAs bind to their RNA target with a greater degree of complementarity compared with other organisms. For the reasons already given, DNA oligonucleotides complementary for specific plant miRNA will bind with high specificity to the cDNA template, a factor that will by itself exclude the false priming of the primer and reduce the non-specific products.

One problem of the PCR strategy is the necessity to sequence many clones and to analyse the resulting large volume of data, especially if one applies this method to larger genomes like the human one. However, it should be possible to make use of microarray technology to identify for a specific miRNA a large number of target mRNAs at once. In order to apply microarray technology one has to design for the starting PCR a reverse primer that will add to the cDNA products a phage promoter (T7, SP6 or T3). After the PCR, all products are in vitro transcribed, labelled and applied on a microarray chip loaded with the relevant genome.

The PCR strategy of target identification relies on principles similar to those of the bioinformatics approach. It scans for miRNA binding sites on the basis of the requirement for strong binding interaction at the 3′ end of the primer (corresponding to the 5′ end of the miRNA), plus some additional interaction at the 5′ end of the oligonucleotide. The advantage of using this method versus the computational approach is that the positive outcome is a result of a real base interaction and not a hypothetical, predicted base-pairing. With the PCR method on a cDNA library one also screens for binding motifs that are located within the coding sequence. This is in contrast to most computer algorithms that focus only on 3′ UTR sequences and would not detect miRNA binding sites within coding regions as they have been found in plants. A further advantage of the experimental strategy is that it focuses on genes that are actually expressed in a specific tissue or at a certain developmental stage, depending on the starting material used for the cDNA library. Alternatively, the method can be used directly on purified mRNAs using oligo dT as primer for RT, followed by PCR with the anti-miDNA oligonucleotide. In either way, different transcriptomes can be screened separately. By contrast, the bioinformatics approach will typically screen all genes in a genome, even those that might not be expressed at a given time and in a given tissue.

4
Validation of miRNA/mRNA Target Pairs by Sensor Constructs

The PCR approach is a powerful method for identifying candidate target genes that are under the control of a miRNA. However, regardless of whether the candidate gene is identified by bioinformatics or by this experimental strategy, functional validation of the interaction is required.

A way to test the miRNA:mRNA interaction in vivo is by specific reporter constructs, also named sensor or indicator constructs (Fig. 1c). These are recombinant mRNAs that are expressed from constitutive promoters and contain the coding region of the reporter and a 3' UTR that contains the binding motif originating from the miRNA-regulated target gene. The basic idea is to monitor the changes in the expression of a reporter protein caused by the binding of the miRNA on the 3' UTR. There are several reports that describe the use of these reporter constructs, both transiently in *Drosophila* Schneider 2 (S2) (Boutla et al. 2003) and HeLa cell cultures (Meister et al. 2004) and as stable transformants in *Drosophila* (Brennecke et al. 2003; Stark et al. 2003).

The 3' UTR can be designed to contain several miRNA binding sites (Brennecke et al. 2003; Meister et al. 2004; Stark et al. 2003) or the authentic 3' UTR of a mRNA target gene may be used (Boutla et al. 2003; Brennecke and Cohen 2003). Changes in the expression of the reporter protein reveal the regulatory pattern generated by the miRNA. If the sensor construct is expressed in a transgenic organism, a spatial pattern may be observed (Brennecke and Cohen 2003), which can be considered as the "negative" of miRNA expression. Mutations within the miRNA binding sites that impair the regulation provide additional proof for miRNA regulation.

Sensor constructs may also be used for quantitative analysis. Boutla et al. (2003) engineered a sensor construct with the authentic 3' UTR of a potential *Drosophila* target gene regulated by miR-13a. They introduced mutations into the prospected miRNA binding site that made the miRNA interaction either stronger or weaker. Mutations that decreased the interaction via the *K*-box motif decreased the binding of the miRNA. As a consequence the levels of the reporter protein were increased almost 2 times compared with the levels in the wild type. On the other hand, mutations resulting in perfect complementarity inhibited strongly the expression of the reporter gene to levels close to 60% of the wild-type expression. However, in this case it was not clear whether the inhibition was due to stronger binding of the miRNA or whether the miRNA operated as an siRNA resulting in target cleavage as observed in other cases (Doench et al. 2003; Hutvagner and Zamore 2002).

Sensor constructs make use of the endogenous miRNA levels. This might be a problem if the sensor concentration is too high. Here, the sensor is expressed, regardless of whether it contains a functional miRNA binding site, because there are not sufficient miRNA molecules to saturate the binding site, so that no negative regulation is possible. In accordance with this, we saw that the relative sensitivity towards the sensor increases if less sensor is used (Boutla et al. 2003).

5
Concluding Remarks

Identifying mRNA:miRNA interactions will be of importance to elucidate miRNA function. Given the many miRNA genes, of which many will have multiple targets, this is a considerable challenge. In addition, one particular mRNA can also be controlled jointly and simultaneously by different miRNAs, which adds to the complexity of translational regulation. The experimental strategies that we presented here mark the first steps towards understanding basic mechanisms by which miRNAs control and modulate gene expression.

References

Abbott AL (2003) Heterochronic genes. Curr Biol 13:R824–825

Altschul SF, Madden TL, Schaffer AA, Zhang J, Zhang Z, Miller W, Lipman DJ (1997) Gapped BLAST and PSI-BLAST: a new generation of protein database search programs. Nucleic Acids Res 25:3389–3402

Ambros V (1989) A hierarchy of regulatory genes controls a larva-to-adult developmental switch in C. elegans. Cell 57:49–57

Ambros V, Horvitz HR (1984) Heterochronic mutants of the nematode Caenorhabditis elegans. Science 226:409–416

Ambros V, Moss EG (1994) Heterochronic genes and the temporal control of C. elegans development. Trends Genet 10:123–127

Ambros V, Bartel B, Bartel DP, Burge CB, Carrington JC, Chen X, Dreyfuss G, Eddy SR, Griffiths-Jones S, Marshall M, Matzke M, Ruvkun G, Tuschl T (2003) A uniform system for microRNA annotation. RNA 9:277–279

Ambros V, Lee RC, Lavanway A, Williams PT, Jewell D (2003) MicroRNAs and other tiny endogenous RNAs in C. elegans Curr Biol 13:807–818

Aravin AA, Lagos-Quintana M, Yalcin A, Zavolan M, Marks D, Snyder B, Gaasterland T, Meyer J, Tuschl T (2003) The small RNA profile during Drosophila melanogaster development. Dev Cell 5:337–350

Boutla A, Delidakis C, Tabler M (2003) Developmental defects by antisense-mediated inactivation of micro-RNAs 2 and 13 in Drosophila and the identification of putative target genes. Nucleic Acids Res 31:4973–4980

Brennecke J, Cohen SM (2003) Towards a complete description of the microRNA complement of animal genomes. Genome Biol 4:228

Brennecke J, Hipfner DR, Stark A, Russell RB, Cohen SM (2003) bantam encodes a developmentally regulated microRNA that controls cell proliferation and regulates the proapoptotic gene hid in Drosophila. Cell 113:25–36

Calin GA, Dumitru CD, Shimizu M, Bichi R, Zupo S, Noch E, Aldler H, Rattan S, Keating M, Rai K, Rassenti L, Kipps T, Negrini M, Bullrich F, Croce CM (2002) Frequent deletions and down-regulation of micro-RNA genes miR15 and miR16 at 13q14 in chronic lymphocytic leukemia. Proc Natl Acad Sci USA 99:15524–15529

Calin GA, Liu CG, Sevignani C, Ferracin M, Felli N, Dumitru CD, Shimizu M, Cimmino A, Zupo S, Dono M, Dell'Aquila ML, Alder H, Rassenti L, Kipps TJ, Bullrich F, Negrini M, Croce CM (2004a) MicroRNA profiling reveals distinct signatures in B cell chronic lymphocytic leukemias. Proc Natl Acad Sci USA 101:11755–11760

Calin GA, Sevignani C, Dumitru CD, Hyslop T, Noch E, Yendamuri S, Shimizu M, Rattan S, Bullrich F, Negrini M, Croce CM (2004b) Human microRNA genes are frequently located at fragile sites and genomic regions involved in cancers. Proc Natl Acad Sci USA 101:2999–3004

Doench JG, Sharp PA (2004) Specificity of microRNA target selection in translational repression. Genes Dev 18:504–511

Doench JG, Petersen CP, Sharp PA (2003) siRNAs can function as miRNAs. Genes Dev 17:438–442

Enright AJ, John B, Gaul U, Tuschl T, Sander C, Marks DS (2003) MicroRNA targets in Drosophila. Genome Biol 5:R1

Grad Y, Aach J, Hayes GD, Reinhart BJ, Church GM, Ruvkun G, Kim J (2003) Computational and experimental identification of C. elegans microRNAs. Mol Cell 11:1253–1263

Griffiths-Jones S (2004) The microRNA registry. Nucl Acids Res 32:D109–D111

Hutvagner G, Zamore PD (2002) A microRNA in a multiple-turnover RNAi enzyme complex. Science 297:2056–2060

Hutvagner G, Simard MJ, Mello CC, Zamore PD (2004) Sequence-specific inhibition of small RNA function. PLoS Biol 2:e98

John B, Enright AJ, Aravin A, Tuschl T, Sander C, Marks DS (2004) Human microRNA targets PLoS Biol 2:e363

Jones-Rhoades MW, Bartel DP (2004) Computational identification of plant microRNAs and their targets, including a stress-induced miRNA. Mol Cell 14:787–799

Juarez MT, Kui JS, Thomas J, Heller BA, Timmermans MC (2004) MicroRNA-mediated repression of rolled leaf1 specifies maize leaf polarity. Nature 428:84–88

Kiriakidou M, Nelson PT, Kouranov A, Fitziev P, Bouyioukos C, Mourelatos Z, Hatzigeorgiou A (2004) A combined computational-experimental approach predicts human microRNA targets. Genes Dev 18:1165–1178

Lagos-Quintana M, Rauhut R, Lendeckel W, Tuschl T (2001) Identification of novel genes coding for small expressed RNAs. Science 294:853–858

Lagos-Quintana M, Rauhut R, Yalcin A, Meyer J, Lendeckel W, Tuschl T (2002) Identification of tissue-specific microRNAs from mouse. Curr Biol 12:735–739

Lagos-Quintana M, Rauhut R, Meyer J, Borkhardt A, Tuschl T (2003) New microRNAs from mouse and human. RNA 9:175–179

Lai EC (2002) Micro RNAs are complementary to 3′ UTR sequence motifs that mediate negative post-transcriptional regulation. Nat Genet 30:363–364

Lai EC, Posakony JW (1997) The bearded box, a novel 3′ UTR sequence motif, mediates negative post-transcriptional regulation of bearded and enhancer of split complex gene expression. Development 124:4847–4856

Lai EC, Burks C, Posakony JW (1998) The K box, a conserved 3′ UTR sequence motif, negatively regulates accumulation of enhancer of split complex transcripts. Development 125:4077–4088

Lai EC Tmancak P, Williams RW, Rubin GM (2003) Computational identification of Drosophila microRNA genes. Genome Biol 4:R42

Lai EC, Wiel C, Rubin GM (2004) Complementary miRNA pairs suggest a regulatory role for miRNA:miRNA duplexes. Rna 10:171–175

Lau NC, Lim LP, Weinstein EG, Bartel DP (2001) An abundant class of tiny RNAs with probable regulatory roles in Caenorhabditis elegans. Science 294:858–862

Lee RC, Ambros V (2001) An extensive class of small RNAs in Caenorhabditis elegans. Science 294:862–864

Lee RC, Feinbaum RL and Ambros V (1993) The C. elegans heterochronic gene lin-4 encodes small RNAs with antisense complementarity to lin-14. Cell 75:843–854

Lee Y, Jeon K, Lee JT, Kim S, Kim VN (2002) MicroRNA maturation: stepwise processing and subcellular localization. EMBO J 21:4663–4670

Lee Y, Ahn C, Han J, Choi H, Kim J, Yim J, Lee J, Provost P, Radmark O, Kim S, Kim VN (2003) The nuclear RNase III Drosha initiates microRNA processing. Nature 425:415–419

Lee Y, Kim M, Han J, Yeom KH, Lee S, Baek SH, Kim VN (2004) MicroRNA genes are transcribed by RNA polymerase II. EMBO J 23:4051–4060

Lee YS, Nakahara K, Pham JW, Kim K, He Z, Sontheimer EJ, Carthew RW (2004) Distinct roles for Drosophila Dicer-1 and Dicer-2 in the siRNA/miRNA silencing pathways. Cell 117:69–81

Lewis BP, Shih IH, Jones-Rhoades MW, Bartel DP, Burge CB (2003) Prediction of mammalian microRNA targets. Cell 115:787–798

Lim LP, Lau NC, Weinstein EG, Abdelhakim A, Yekta S, Rhoades MW, Burge CB, Bartel DP (2003) The microRNAs of Caenorhabditis elegans. Genes Dev 17:991–1008

Lund E, Guttinger S, Calado A, Dahlberg JE, Kutay U (2004) Nuclear export of microRNA precursors. Science 303:95–98

Mallory AC, Vaucheret H (2004) MicroRNAs: something important between the genes. Curr Opin Plant Biol 7:120–125

Meister G, Landthaler M, Dorsett Y, Tuschl T (2004) Sequence-specific inhibition of microRNA- and siRNA-induced RNA silencing. RNA 10:544–550

Olsen PH, Ambros V (1999) The lin-4 regulatory RNA controls developmental timing in Caenorhabditis elegans by blocking LIN-14 protein synthesis after the initiation of translation. Dev Biol 216:671–680

Papp I, Mette MF, Aufsatz W, Daxinger L, Schauer SE, Ray A, van der Winden J, Matzke M, Matzke AJ (2003) Evidence for nuclear processing of plant micro RNA and short interfering RNA precursors. Plant Physiol 132:1382–1390

Park W, Li J, Song R, Messing J, Chen X (2002) CARPEL FACTORY, a Dicer homolog, and HEN1, a novel protein, act in microRNA metabolism in Arabidopsis thaliana. Curr Biol 12:1484–1495

Pasquinelli AE, Reinhart BJ, Slack F, Martindale MQ, Kuroda MI, Maller B, Hayward DC, Ball EE, Degnan B, Muller P, Spring J, Srinivasan A, Fishman M, Finnerty J, Corbo J, Levine M, Leahy P, Davidson E, Ruvkun G (2000) Conservation of the sequence and temporal expression of let-7 heterochronic regulatory RNA. Nature 408:86–89

Rajewsky N, Socci ND (2004) Computational identification of microRNA targets. Dev Biol 267:529–535

Reinhart BJ, Slack FJ, Basson M, Pasquinelli AE, Bettinger JC, Rougvie AE, Horvitz HR, Ruvkun G (2000) The 21-nucleotide let-7 RNA regulates developmental timing in Caenorhabditis elegans. Nature 403:901–906

Reinhart BJ, Weinstein EG, Rhoades MW, Bartel B, Bartel DP (2002) MicroRNAs in plants. Genes Dev 16:1616–1626

Rhoades MW, Reinhart BJ, Lim LP, Burge CB, Bartel B, Bartel DP (2002) Prediction of plant microRNA targets. Cell 110:513–520

Schauer SE, Jacobsen SE, Meinke DW, Ray A (2002) DICER-LIKE1: blind men and elephants in Arabidopsis development. Trends Plant Sci 7:487–491

Slack FJ, Basson M, Liu Z, Ambros V, Horvitz HR, Ruvkun G (2000) The lin-41 RBCC gene acts in the C. elegans heterochronic pathway between the let-7 regulatory RNA and the LIN-29 transcription factor. Mol Cell 5:659–669

Stark A, Brennecke J, Russell RB, Cohen SM (2003) Identification of Drosophila MicroRNA targets. PLoS Biol 1:e60

Vella MC, Choi EY, Lin SY, Reinert K, Slack FJ (2004) The C. elegans microRNA let-7 binds to imperfect let-7 complementary sites from the lin-41 3′ UTR. Genes Dev 18:132–137

Wang JF, Zhou H, Chen YQ, Luo QJ, Qu LH (2004) Identification of 20 microRNAs from Oryza sativa. Nucleic Acids Res 32:1688–1695

Xie Z, Johansen LK, Gustafson AM, Kasschau KD, Lellis AD, Zilberman D, Jacobsen SE, Carrington JC (2004) Genetic and functional diversification of small RNA pathways in plants. PLoS Biol 2:E104

Xu P, Vernooy SY, Guo M, Hay BA (2003) The Drosophila microRNA Mir-14 suppresses cell death and is required for normal fat metabolism. Curr Biol 13:790–795

Nucleic Acids and Molecular Biology, Vol. 17
Wolfgang Nellen, Christian Hammann (Eds.)
Small RNAs
© Springer-Verlag Berlin Heidelberg 2005

Protein Interactions
with Double-Stranded RNA in Eukaryotic Cells

Christian Hammann

AG Molecular Interactions, Department of Genetics, University of Kassel,
Heinrich-Plett-Strasse 40, 34132 Kassel, Germany
c.hammann@uni-kassel.de

Abstract Double-stranded RNA has long been known to be a trigger for cellular responses
to viral infections, leading to dramatic changes in cellular processes. Since the advent
of RNA interference, it has become clear that double-stranded RNA also causes specific effects, regulating gene expression on the transcriptional, post-transcriptional and
translational levels. An essential prerequisite for double-stranded RNA effects is proteins
that specifically recognise these molecules in order to elicit the cellular response. This
chapter focuses on the function and molecular architecture of those proteins that interact with double-stranded RNA and that are key players in the RNA interference, editing
and the PKR response. After summarising the origin of double-stranded RNA molecules
and structural features of A-type helices, the way proteins can interact with this secondary structure is discussed. The variability of domain structures of proteins that are
functional homologues in processes triggered by double-stranded RNA is reviewed and
consequences resulting from the different design of proteins from various organisms are
discussed. Finally, differences and similarities of pathways with respect to their subcellular localisation and the length of the double-stranded RNA trigger are summarised.

1
Introduction

The central dogma of molecular biology, DNA makes RNA makes protein,
considered RNA molecules as mere vehicles en route to protein biosynthesis. This applied both to messenger RNA (mRNA) and to "classical" noncoding RNAs, transfer and ribosomal RNAs, all of which are formally single-
stranded. Double-stranded RNA was considered to be absent from cells, with
the exception of viral RNA that cells would act upon by various counterdefence mechanisms.

These views have clearly changed, initially by the discovery of several endogenous antisense RNAs and more recently, as well as more profoundly, by
the discovery of RNA interference (RNAi; Fire et al. 1998). Double-stranded
RNAs of variable origin play central roles in the specific cellular regulation at
the transcriptional, post-transcriptional and translational levels. The decision
as to which of the pathways is triggered by double-stranded RNA depends
on their origin, their cellular localisation and their length. These regulatory
pathways also partially overlap with the less-specific classical antiviral de-

fence mechanisms. This could be attributed to the fact that the majority of proteins that interact with double-stranded RNA do so by virtue of the same protein domain, the double-stranded RNA binding domain (dsRBD). Indeed, all dsRBDs have the capacity to interact with double-stranded RNA, and this domain is therefore often considered to be non-sequence-specific. However, several recent reports indicate that individual dsRBDs might have, on top of a general binding capacity, intrinsic preferences for certain sequence or structure elements within or adjacent to double-stranded stretches of RNA.

2
Occurrence and Origin of Cellular Double-Stranded RNA

RNAs with the potential to form double strands are found in all kingdoms of life. Whether a double strand is formed depends on the subcellular localisation within the appropriate organism, and on the nature of the transcript(s) involved. Double strands can originate from natural antisense transcripts (NATs), which might be encoded either in the opposite DNA strand at the same locus (cis-NATs) or at a separate locus (trans-NATs; Lavorgna et al. 2004). Furthermore, in many organisms, particularly plants (Gazzani et al. 2004), and also, e.g. *Neurospora crassa* (Chicas et al. 2004) or *Dictyostelium discoideum* (Martens et al. 2002), but not mammals or *Drosophila melanogaster* (Stein et al. 2003), RNA-dependent RNA polymerases (RdRPs) are present, which can transcribe an antisense RNA from an RNA template. RdRPs are also central to the replication and infection of (+) strand RNA viruses; however, this extensive field has recently been reviewed elsewhere (van Dijk et al. 2004) and will not be discussed here.

Imperfect RNA double strands are found in the primary transcripts of microRNAs (pri-miRNAs; Ambros 2004) and can, in principle, occur within any RNA transcript, as exemplified by the intron–exon junction of the precursor mRNA (pre-mRNA) of the glutamate receptor subunit B pre-mRNA (GluR-B; Melcher et al. 1996). Recent data indicated that perfect RNA double strands might also be formed intramolecularly within certain endogenous transcripts (Bartsch et al.. 2004; Gräf et al. 2004) by virtue of conventional Watson–Crick base-pairing. However, whether this potential of double-strand formation is realised in the cell is unknown at present. The formation of this secondary structure could be prevented either by kinetic traps during transcription or, more actively, altered by virtue of RNA chaperones (Schroeder et al. 2004).

2.1
Database Searches for Antisense RNA and Experimental Validation

In recent years, several database searches for NATs, particularly from mammalian organisms, and for intramolecular double strands in endogenous

RNAs have been carried out. In one of the first such searches, Sanderson and co-workers predicted more than 800 antisense transcripts in the human transcriptome (Lehner et al. 2002). Subsequently, this number was increased considerably, on the basis of more sophisticated database searches and, importantly, from experiments validating the expression of appropriate antisense RNAs (Chen et al. 2004; Røsok and Sioud 2004; Yelin et al. 2003). While the numerical estimates for appropriate transcripts vary between 5 and 20%, it is clear now that endogenous antisense RNA is a common and important feature of mammalian cells. However, this observation is not limited to mammals, as similar values have been reported for *Drosophila* (Misra et al. 2002) and predicted for the highly compact genome of *Fugu rubripes* (Dahary et al. 2005). In *Arabidopsis thaliana*, NATs seem to be even more abundant (Stolc et al. 2005; Yamada et al. 2003). NATs can, a priori, form double strands with the respective sense RNA by virtue of conventional Watson–Crick base-pairing; however, it is not clear whether they do (Munroe 2004). One of the best studied sense–antisense pairs is the bacterial CopA–CopT. Despite perfect base complementarity, hybridisation of the two RNA molecules does not result in a full double strand, but rather in a complex four-way junction (Kolb et al. 2000). Also, at least for mammals, formation of extended double strands by endogenous sense and antisense transcripts seems unlikely, as this would be expected to lead to cell death mediated by the PKR pathway (see later).

Next to NATs, self-complementary sequences have also been found that can form double strands within individual RNA molecules. For *Dictyostelium*, a database search identified such sequences; however, not within mRNAs, but exclusively in antisense orientation to mRNAs (Gräf et al. 2004). Intriguingly, at least one of these antisense RNAs is expressed and regulates the stability of the complementary psvA mRNA during *Dictyostelium* development (Hildebrandt and Nellen 1992). While it is not clear whether further antisense RNAs are expressed in *Dictyostelium*, there seems to be a selection against the occurrence of double-stranded regions within mRNAs in this organism. At present, there are no similar data available for other organisms, and it would be interesting to see whether the lack of intramolecular double strands in mRNAs is evolutionarily connected to the presence of an RNAi machinery. In this case, organisms lacking RNAi, such as *Saccharomyces cerevisiae* (Aravind et al. 2000), could be expected to contain complementary sequences within mRNAs, allowing for the formation of intramolecular double strands that would not cause a cellular response.

Further examples of sequences that can form double strands within RNA molecules are hybrid RNAs that are not contiguously encoded in the genome, but rather consist of sense and antisense parts (Bartsch et al. 2004). As a source for such mammalian transcripts, a functional RdRP has been postulated by these authors and others (Røsok and Sioud 2004). At present, however, no homologous sequences to known RdRPs have been found in mammalian genomes. If such enzymes existed in mammals, they presumably would represent a novel class of proteins, distinct from the documented RdRPs and their conserved domains.

Overall, these results indicate that complementary RNA sequences occur much more frequently in the cell than previously anticipated. Independent of whether such complementary sequences occur intramolecularly or intermolecularly, they all have the potential to form double strands. The consequences of the formation of double-stranded RNA vary greatly for different eukaryotes, and additionally depend on the subcellular localisation of the RNA(s) and the length of the double-stranded region. Before discussing this, it is instructive to review some structural features of double-stranded RNA, as they greatly influence the interaction with double stranded RNA binding proteins that are responsible for the cellular effects.

2.2
Structural Features of Double-Stranded RNA

The standard helical structure of double-stranded RNA is clearly distinct from that of DNA, and it is this topological difference that makes up for a different mode of recognition by double stranded RNA binding proteins. B-DNA shows a narrow minor groove and a wide major groove through which proteins can interact with the DNA in a sequence-specific manner. The additional 2′-hydroxyl group of RNA does not allow for the formation of a B-type helix owing to steric collision. Instead, double-stranded RNA forms the broader A-type helix, in which the minor groove is wider and the major groove is considerably narrower than in B-type helices (Fig. 1). The narrow major groove in A-type helices, which is also termed a deep groove, renders them essentially inaccessible for proteins, largely preventing sequence-specific interaction with double-stranded RNA. Both these well-known standard structures of double-stranded nucleic acids, B-type DNA and A-type RNA are right-handed. While double-stranded RNA is not found as B-type helices, it can adopt, in exceptional cases, a left-handed Z-type helix (Hall et al. 1984), a structure originally found in DNA under high salt conditions (Wang et al. 1979). Unlike A-type and B-type helices, helices of this type cannot be formed by any two strands of complementary nucleotides, but rather are restricted to alternating purine–pyrimidine sequences. Owing to alternating sugar puckers (3′-endo for purines in syn and 2′-endo for pyrimidines in anti), the phosphodiester backbone of this unusual double helix is not contiguous but runs in zigzags, after which the Z-type helix is named.

There is convincing evidence for the cellular occurrence of Z-DNA, and a Z-DNA binding protein domain has been defined and also functionally characterised (summarised in Rich 2004). This domain, named Zα, is found, for example, in the enzyme ADAR1, an adenosine deaminase acting on RNA (Schwartz et al. 1999). The Z-DNA binding domain also binds Z-RNA (Brown et al. 2000), and there is an early paper reporting the presence Z-RNA in the

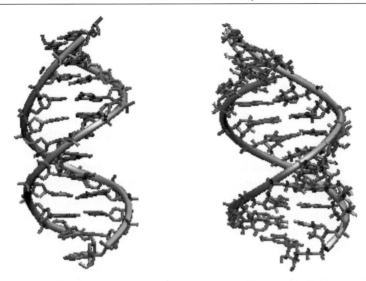

Fig. 1 Helical structures in double-stranded nucleic acids. A comparison of B-type helical double-stranded DNA (*dsDNA*, *left*) and A-type helical double-stranded RNA (*dsRNA*, *right*) is shown. B-type dsDNA has a narrow minor groove and a wide major groove that allows for sequence-specific interactions with respective proteins. This is not possible in A-type dsRNA as the major groove is too narrow. Most dsRNA binding proteins therefore interact with the nucleic acid in a sequence-independent manner. The figure was created with VMD software (Humphrey et al. 1996) using coordinates for the dsDNA from X-ray analysis (Dickerson dodecamer; Drew et al. 1981, PDB accession 1BNA) and for the dsRNA from NMR analysis (Conte et al. 1997, PDB accession 1AL5)

cytosol (Zarling et al. 1987). Despite this, the biological significance of Z-RNA is under debate, and the similar structures of Z-RNA and Z-DNA might be the actual reason for the observed interaction of Z-RNA with the Z-DNA binding domain. This situation is clearly distinct from that of conventional A-type and B-type helices of RNA and DNA, respectively, where, by virtue of the mode of interaction, proteins bind selectively one type of nucleic acid.

Beautiful crystallographic work on the ribosome has highlighted the great structural variability of RNA molecules, with conventional Watson–Crick base-pairing accounting for just about 60% of the observed RNA–RNA interactions (Ban et al. 2000; Schluenzen et al. 2000; Wimberly et al. 2000). From this and other RNA structures solved at high resolution, an exhaustive description and concise nomenclature of possible base-pair interactions in RNA have been developed (Leontis and Westhof 2001). Whilst most of the non-canonical base pairs are involved in tertiary contacts, it should be pointed out that base pairs isosteric to Watson–Crick pairs can occur in double-stranded RNA without altering its A-type helical structure substantially. This further increases the sequence space for the formation of natural RNA double strands in the cell, compared with that of double-stranded DNA.

3
Protein Interactions

3.1
Protein Domains Interacting with Double-Stranded RNA

There are a number of identified protein domains that interact with RNA. These domains include short motifs of about 10–25 amino acids, like the K-homology (KH) domain, the arginine-rich motif (ARM), the arginine–glycine–glycine (RGG) box and the oligonucleotide/oligosaccharide binding (OB) fold. Larger RNA-binding protein domains of 50–100 amino acids include the RNA recognition motif (RRM), the dsRBD, and the CCHH-type zinc-finger domain, which also binds double-stranded DNA (reviewed in Draper 1999; Hall 2002; Messias and Sattler 2004; Varani and Nagai 1998). Among these, since they have been studied the best, the RRM serves as a paradigm of single-stranded RNA binding domains and the dsRBD for double-stranded binding domains. For both these domains, excellent reviews have been published recently (Hall 2002; Saunders and Barber 2003; Tian et al. 2004) and in the context of this article, only the key features of the dsRBD will be discussed.

The dsRBD, also termed the double-stranded RNA binding motif (dsRBM), was originally discovered in the *Drosophila* Staufen protein and in the *Xenopus laevis* Xlrbpa protein (St Johnston et al. 1992). The domain is made up by about 70 amino acids and its interaction with double-stranded RNA has been postulated to be sequence-independent. This was concluded from the observation that dsRBDs will bind to any double-stranded RNA, for example in gel retention assays, and is in line with the structural features of A-type helices laid out earlier. Consequently, neither DNA nor single-stranded RNA can be bound (Bass et al. 1994; St Johnston et al. 1992). After their identification, dsRBDs were subsequently found to occur in a wide range of different proteins with functions in RNAi, RNA localisation, RNA processing and editing, as well as translational repression (Beal 2005; Saunders and Barber 2003; Stefl et al. 2005; Tian et al. 2004). In many of these proteins, more than one copy of the dsRBD is found and the copy number is of great importance for the respective protein functions (Sect. 3.2). In light of its wide occurrence, both NMR and X-ray crystallography have been applied to determine the structure of the domain (Bycroft et al. 1995; Kharrat et al. 1995; Nanduri et al. 1998; Ramos et al. 2000; Ryter and Schultz 1998; Wu et al. 2004). From these studies, a unifying picture emerged in which the dsRBD adopts an $\alpha\beta\beta\beta\alpha$ topology with the two α-helices arranged next to each other on one face of a three-stranded antiparallel β-sheet (Fig. 2). In both, RNA-free and RNA-containing structures, the amino acids that have been implied in mediating contacts to the nucleic acid are oriented such that they face one edge of the dsRBD. The respective interacting amino acids are located within the two α-helices and within a loop that connects two strands of the β-sheet (loop $\beta1–\beta2$; Fig. 2a).

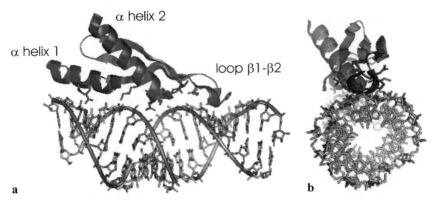

α helix 2

α helix 1

loop β1-β2

a   b

Fig. 2 Binding of the dsRNA binding domain (*dsRBD*) to dsRNA. **a** The dsRBD of Xlrbpa bound to dsRNA is shown, using coordinates (PDB accession 1DI2) from X-ray crystallographic work (Ryter and Schultz 1998). The dsRBD adopts an $\alpha\beta\beta\beta\alpha$ fold, in which amino acids of both α-helices and of the loop connecting strands 1 and 2 of the β-sheet, loop $\beta1$–$\beta2$, make contacts to the dsRNA. The majority of contacts are sequence-independent, and rather recognise the shape of the A-type double-stranded helix. **b** The dsRBD contacts one face of the dsRNA only. The figure was created using Pymol software (DeLano 2002)

From the crystal structure of the Xlrbpa dsRBD bound to an RNA duplex of ten base pairs (Ryter and Schultz 1998), several general features can be deduced. A single dsRBD binds to more than a full turn of the A-type helix, making contacts with two minor grooves and the major groove in between, thereby spanning about 16 base pairs (Fig. 2a). The binding of the dsRBD to double-stranded RNA apparently results only in local structural changes, as the conformation of the dsRBD with RNA bound differs only slightly from that of the RNA-free form that was solved earlier by NMR studies. Also the RNA A-form helix is only very slightly distorted in complex with the protein domain. Further, this crystal structure shows that with an interacting amino acid residue oriented on one face of the domain, only one face of the RNA double strand is recognised (Fig. 2b). The interactions themselves are mainly to 2′ hydroxyls (minor groove interactions) and to phosphate backbone and/or non-bridging oxygens (major groove interactions). While this is in line with the idea of sequence-independent interactions, surprisingly, contacts to base functional groups were also observed in the crystal structure, which often, but not always, were water-mediated. For water-mediated contacts it can be easily envisaged that the interacting amino acids would adjust their minor groove hydrogen binding pattern to any sequence. However, such a modulation seems less likely for direct contacts, as seen in the Xlrbpa dsRBD for Pro140, which interacts with the exocyclic amino group of a guanosine, and thus seems to be specific for a GC pair at this position (Ryter and Schultz 1998).

While the nature of the interactions between dsRBDs and their target RNA thus, in general, is in agreement with the sequence-independent binding, it cannot be ruled out that amino acids at respective positions in individual dsRBDs might confer some degree of preferential, possibly even specific binding. This could be achieved by interactions with preferred sequences, as seen in the Xlrbpa structure, or with preferred topologies, as exemplified by the NMR structure of the dsRBD of the RNase III RntpI from *S. cerevisiae* (Wu et al. 2004), in which the shape of a tetraloop is recognised by the α-helix 1. Furthermore, the majority of target sequences show structural deviations from A-type helices, by the presence of either bulges or loops—yet these sequences are natural targets of the dsRBDs in important proteins, such as PKR, Dicer or ADAR.

Several recent studies support a sequence-related or a structure-related preferential binding by dsRBDs. While the binding behaviour of the dsRBD was originally described on the basis of bulk experiments, these new studies used either methods investigating the interaction on a single-molecule level or mapped the binding in footprinting assays, and thus allowed for a much more detailed view on the mode of interaction. In the first such study, scanning force microscopy was applied to the RNA editing enzyme ADAR2 (Sect. 3.2.3), which contains two consecutive dsRBDs, and one of its natural targets, the R/G within the pre-mRNA of GluR-B (Klaue et al. 2003). For analysis of this interaction on a single-molecule level, the target RNA was embedded in an RNA backbone molecule containing a large double-stranded central part. Editing was observed predominantly at the R/G site, but also promiscuously in the double strand. Intriguingly, this study also identified distinct binding sites that would not be edited by ADAR2, and, using sequence variants, allowed for the discrimination of binding and editing events. In a second study investigating ADAR2, binding of its dsRBDs to short model substrates mimicking the Q/R site of the same pre-mRNA was studied by directed hydroxyl radical cleavage analysis, and compared with the binding of a different dsRBD from the PKR (Stephens et al. 2004). As expected for dsRBDs, both domains interacted with the model substrate; however, they showed a clear preference as they bound at different positions in the (imperfect) double strand. This is not to say that dsRBDs are the only determinant of target selection within a given double-stranded RNA and likely other protein domains next to the dsRBDs contribute to sequence specificity or preference. Yet, these examples show that individual dsRBDs have intrinsic preferences that allow us to define, in concert with the other protein domains, specific double-stranded RNA targets of the enzymes in which they are contained. Furthermore, there are proteins with specific target RNAs whose only protein domains are dsRBDs (Sect. 3.2.1), yet not all of them bind double-stranded RNA strongly, like in the case of PACT or Staufen (Micklem et al. 2000; Peters et al. 2001). These observations argue additionally against a merely structure-dependent interaction of dsRBDs with double-stranded RNA.

While the understanding of the interaction of dsRBDs with RNA helices is thus deepening, it is likely that other, less well-characterised protein domains can also mediate a protein interaction with RNA double-stranded structures. The $2',5'$-oligoadenylate synthetase (2,5-AS), for example, is directly activated by the presence of double-stranded RNA as an antiviral defence mechanism, yet all isoforms of the enzyme are devoid of a dsRBD or another well-defined RNA binding domain. Nucleic acid binding is, in this case, attributed to the PAP and 2,5-AS domains; however, further experimental evidence is required to clarify the details of this presumed interaction.

The amazing revolution that the field has undergone since Fire et al. (1998) published their paper on RNAi in *Caenorhabditis elegans* led to the identification of PAZ as a further protein domain with the potential to bind RNA, by preferentially interacting with the two-nucleotide $3'$ overhangs of double-stranded small interfering RNAs (siRNAs) (Lingel et al. 2003; Ma et al. 2004; Song et al. 2003; Yan et al. 2003; Zhang et al. 2004). This PAZ domain, named after the *Drosophila* proteins Piwi, Argonaute and Zwille, occurs in several proteins that have essential functions in mechanisms of post-transcriptional gene silencing (PTGS), like Ago2 and members of the Dicer family of RNase III proteins (Carmell et al. 2002; Cerutti et al. 2000).

3.2
Molecular Architecture of dsRBD-Containing Proteins

Proteins with dsRBDs have been found in all eukaryotic organisms investigated, and they can be divided into two classes, those that contain additional protein domains and those that do not. In the latter case, the proteins contain between two and five dsRBDs that are often found to cover the entire amino acid sequence. The more dsRBDs these proteins have, the likelier it is that not all of them can interact strongly with double-stranded RNA. The majority of dsRBD-containing proteins show, in contrast to those that were mentioned before, several other functional domains (summarised in Fig. 3). In the following, the different protein classes will be discussed, starting with those that only contain dsRBDs, followed by three selected families with additional protein domains, PKR, ADARs and RNase III-like proteins.

3.2.1
Proteins with dsRBD as the Only Functional Domain

Proteins with no other identified functional domain next to their dsRBDs localise preferentially, but not exclusively, to the cytoplasm, where they play roles in diverse mechanisms, such as translational control, development and RNAi.

Staufen was discovered in *Drosophila*, where it functions at distinct stages of development by anchoring *bicoid* and *oskar* mRNAs to the anterior and

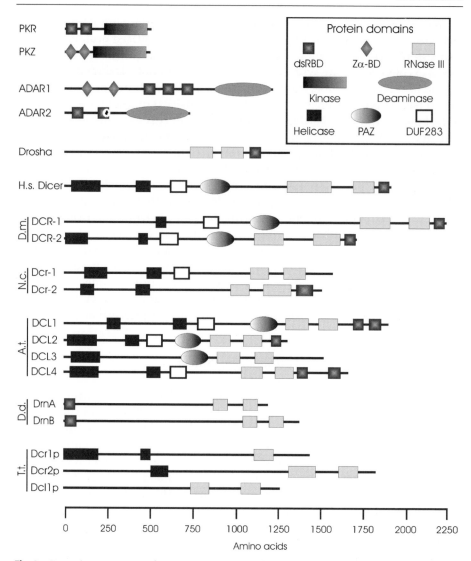

Fig. 3 Domain structure of representative members of protein families with dsRBDs. The proteins are discussed in detail in the text. Protein domains were determined using the protein family database (Pfam) at the InterProScan site of the European Bioinformatics Institute (http://www.ebi.ac.uk/InterProScan/). Protein domains in the *inset* are as follows (Pfam accession numbers and Pfam definition are stated in *parentheses*): dsRBD (PF00035; dsRNA binding), Zα-BD (PF02295; adenosine deaminase Zα domain), RNase III (PF00636; RNase III domain), kinase (PF02137; protein kinase domain); deaminase (PF02137, adenosine-deaminase, editase, domain); helicase (PF00270, DEAD/DEAH box helicase and PF00271, helicase conserved C-terminal domain), PAZ (PF02170; PAZ domain); and DUF283 (PF03368; domain of unknown function). *H.s.* Homo sapiens, *D.m. Drosophila melanogaster, N.c. Neurospora crassa, A.t. Arabidopsis thaliana, D.d. Dictyostelium discoideum, T.t. Tetrahymena termophila*

posterior of the oocyte, respectively (St Johnston et al. 1991). The protein contains formally five dsRBDs; however, only three of them interact strongly with double-stranded RNA, while the other two, dsRBD 2 and dsRBD 5, which are conserved to a lesser extent, are functionally implied in mRNA localisation and activation (Micklem et al. 2000). Homologues of Staufen are also present in mammalian organisms, where they are implied in cellular mRNA transportation and localisation (summarised in Saunders and Barber 2003).

The two dsRBDs of the *Drosophila* R2D2 occupy the N-terminal half of the about 300 amino acid protein. R2D2 plays a central role in RNAi in *Drosophila* (Liu et al. 2003), by stably associating with Dicer-2 (DCR-2; Fig. 3), which is required for processing of long double-stranded RNA in the RNAi pathway in this organism (Lee et al. 2004). It is interesting to note that homologues of R2D2 are largely absent from the database. The RNAi pathways of *Drosophila* and mammals are similar and share many protein components (summarised in Meister and Tuschl 2004). Yet, with respect to long double-stranded RNA, there are differences, as they trigger the PKR pathway in mammals rather than RNAi. However, in cells devoid of the PKR system, RNAi can be triggered by long RNA double strands (Billy et al. 2001). The presence of the PKR pathway in mammalian organisms therefore seems to exclude a functional diversification of Dicer enzymes such as that observed in *Drosophila*, which lacks the PKR response. This view is supported by the absence of not only R2D2 but also DCR-2 homologues in mammals, which contain one Dicer protein only that can, in principle, function in both pathways carried out by the two Drosophila enzymes.

If an R2D2 function was required in other eukaryotic organisms, it is possible that this would be fulfilled by functionally homologous proteins. In *C. elegans* and *A. thaliana*, proteins have been identified that have the same domain structure with two N-terminal dsRBDs and are involved in RNA-mediated gene silencing mechanisms; however, they share only little or no significant amino acid identity with R2D2, respectively. RDE-4, for example, binds to double-stranded RNA exclusively during the initiation steps of the RNAi pathway, where it associates with the only Dicer of *C. elegans* (Tabara et al. 2002). The *A. thaliana* protein HYL1, on the other hand, does not function in PTGS but is involved in the nuclear processing of miRNA precursors that can explain developmental abnormalities caused by mutations of *hyl1* (Han et al. 2004; Vazquez et al. 2004). In line with functioning in the miRNA pathways, HYL1 specifically associates with the miRNA processing Dicer from *A. thaliana* (DCL1, see later); similarly, homologues of HYL1 also interact with specific Dicer-like proteins in *A. thaliana* (Hiraguri et al. 2005), suggesting a modulation of Dicer function.

Next to these proteins involved in RNA-mediated gene silencing, two dsRBD proteins are involved in the mammalian PKR pathway and developmental mechanisms, the cytoplasmic protein activator of PKR (PACT; Patel and Sen 1998) and the cytoplasmic and nuclear TAR RNA binding protein (TRBP; Gatignol et al. 1991). The latter was discovered as a protein involved in HIV-1 activation, and its normal cellular function was unravelled subsequently. Both these proteins contain three dsRBDs each. TRBP is an antagonist of the PKR and can prevent, presumably by competing for double-stranded RNA, the autophosporylation of PKR that is required for its activation (Park et al. 1994). The murine homologue, PRBP, has been implied in translational inhibition that is required for spermatogenesis (Lee et al. 1996). It thus seems that the cellular role of TRBP is in translational control, and that HIV adopted its binding capacities for the regulation of viral expression. PACT, made up of three dsRBDs like TRBP, has the opposite effect on the PKR reaction by activating the kinase. The three dsRBDs cover 62% of the amino acid sequence of PACT, but only two of them bind double-stranded RNA strongly, while the third is implied in PKR activation (Peters et al. 2001). These two proteins show high homology to the *Xenopus* Xlrbpa protein, which also contains three dsRBDs. Xlrbpa seems to serve as a general double-stranded RNA binding module, both in the cytoplasm and in the nucleus, associating with ribosomes and heterogeneous nuclear ribonucleoproteins (Eckmann and Jantsch 1997). Its exact function, however, remains elusive.

3.2.2
The Double-Stranded RNA-Activated Protein Kinase PKR

As a counterdefence to long RNA double strands of viral origin, the interferon-induced protein kinase PKR, formerly named DAI or p68, is activated in mammalian cells. Apart from two dsRBDs the protein contains a conserved kinase domain for catalytic activity (Fig. 3). It was discovered in rabbit reticulocyte lysates that showed drastically reduced levels of protein synthesis due to the presence of low concentrations of double-stranded RNA. The responsible activity was shown to be a kinase that specifically phosphorylated the translation initiation factor eIF2α, thus inhibiting translation at the ribosome (Farrell et al. 1977). Details of the activation of PKR were uncovered subsequently, and it was shown that the process that leads eventually to phosphorylated eIF2α was triggered by the activation of PKR into a autophosphorylated dimer (Galabru and Hovanessian 1987). Importantly, PKR requires long stretches of perfect double-stranded RNA for binding and subsequent activation, with a minimum of 30 base pairs and a maximum activation at 85 base pairs (Manche et al. 1992), which, in general, allows for a discrimination from other pathways that are triggered by smaller RNA double strands. Binding of RNA is selective for certain viral target sequences (Circle et al. 1997; Spanggord et al. 2002), and the two dsRBDs of the PKR presumably bind their target RNAs in a cooperative manner, since they show different affinities, yet the presence of both is required for enzyme activity (Tian and Mathews 2001). PKR has been found in mammalian organisms only; however, a recent report described a related protein from zebrafish (*Danio rerio*), in which the two dsRBDs are replaced by two Zα domains that are implied in binding Z-helical DNA (Rothenburg et al. 2005); Homologues of this protein, PKZ, also seem to exist in other fish, like carp and goldfish. Very similarly to PKR, PKZ was found to be induced after injection of poly(inosinic)–poly(cytidylic) acid and to inhibit translation. It thus would seem that PKZ was a functional homologue of PKR. At present, however, it is not clear whether PKZ has the ability to directly interact with double-stranded RNA (S. Rothenburg, personal communication). Alternatively, the dsRBDs could have been swapped to a different protein that would confer the RNA binding capacity in a complex with PKZ.

3.2.3
Adenosine Deaminases That Act on RNA

Another protein containing the Zα domain is one of the adenosine deaminases that act on RNA (ADAR). The reaction catalysed by ADARs is a deamination of adenosine to inosine, which is translated as guanosine (A → I editing). This deamination is a simple hydrolysis reaction, without need for other substrates, as shown by mass spectrometry using $H_2^{18}O$, which identified water as the source of the oxygen atom at the C6 position in inosine (Polson et al. 1991). Proteins of this family occur in a wide range of eukaryotic organisms, and they are characterised by the presence of two or three dsRBDs and the catalytic deaminase domain. In humans, two genes, ADAR1 and ADAR2, encoding active deaminases of this family were identified (Fig. 3).

There are two major splice variants of ADAR1, the longer of which is shown in Fig. 3. This variant is found in the nucleus and the cytoplasm, contains N-terminally two Zα domains and is expressed from an interferon-inducible promoter (George and Samuel 1999). A shorter splice variant lacking the Zα domains is constitutively expressed and localises predominantly to the nucleus (Patterson and Samuel 1995). The molecular architecture of this shorter ADAR1 splice variant thus resembles that of ADAR2, with one less dsRBD in the latter (Fig. 3). As discussed in Sect. 3.1, the dsRBD of ADAR2 contributes to target sequence selection, and the same holds for ADAR1, as shown by swapping its dsRBDs against those of the PKR, which resulted in dramatically changed editing properties of the hybrid protein (Liu et al. 2000).

The A → I editing reactions carried out by ADARs serve at least two purposes: one is the production of proteins with amino acid sequences that are not encoded in the gene, and the other is (hyper-) editing of RNA of viral origin. Editing of mRNA is often observed in intron-containing sequences, and it is the base-pairing of the intron–exon junction that determines the position of the editing event, placing nuclear editing before or at the time of splicing (Higuchi et al. 1993). Targets of editing are often pre-mRNAs encoding receptors for neurotransmitter proteins, presumably as a way of fine-tuning neuronal activity. One of the best studied pre-mRNAs and likely the most important target pre-mRNA is GluR-B, in which editing takes place at two defined positions, the R/G and Q/R sites, named after the amino acid changes that editing causes. In vitro, the R/G site is edited by both ADAR1 and ADAR2, while the Q/R site is deaminated by ADAR2 only (Melcher et al. 1996). ADAR2 -/- mice show reduced levels of editing and die shortly after birth. Intriguingly, substitution of both GluR-B alleles with mutants encoding the edited version of the Q/R site rescued the phenotype, defining the GluR-B pre-mRNA as the physiologically most important target sequence (Higuchi et al. 2000).

Next to the editing of specific positions in certain mRNAs, ADARs also deaminate promiscuously long double-stranded RNAs (Bass and Weintraub 1988). Such non-selective editing could be an antiviral defence mechanism, as infection with DNA or RNA viruses can result in the presence of long double-stranded RNA. Upon hyper-editing of these RNA molecules, their fate is either association with a nuclear matrix associated protein complex resulting in their nuclear retention (Zhang and Carmichael 2001) or, if they can escape the nucleus, degradation by a cytoplasmic RNase activity ("I-RNase") specific for edited RNAs (Scadden and Smith 2001). On the other hand, hyper-editing of viral RNAs is also beneficial for the virus, as the imperfect RNA double strand is to a lesser extent a trigger for an RNAi or a PKR response. Furthermore, viral sequences also usurp the editing machinery specifically, as observed for hepatitis delta virus (HDV), whose antigenome is selectively edited to convert a stop codon, thereby creating a second, extended open reading frame for the hepatitis delta antigen (Polson et al. 1996).

3.2.4
RNase III-Like Proteins

Eukaryotic RNase III-like proteins such as Dicer and Drosha play a central role in RNA-mediated gene silencing on transcriptional, post-transcriptional and translational levels. These processes are discussed in other contributions in this volume. Here, only the domain architecture of individual RNase III-like enzymes will be discussed, as well as consequences of function, resulting from their different designs.

The protein domains in mammalian Dicers are well conserved. These proteins contain N-terminal helicase domains, annotated as DEAD/DEAH box and C-terminal helicases, a PAZ domain, two RNase III domains and a single dsRBD at the C-terminus (cf. the human Dicer in Fig. 3). Additionally, a central DUF283 domain of unknown function is present between the C-terminal helicase and the PAZ domain. This design is not restricted to mammals but is also present in a few other organisms, like *C. elegans* and *A. gambiae*, which also contain a single Dicer gene only. Several other model organisms like

A. thaliana, D. melanogaster, D. discoideum, N. crassa and *Tetrahymena termophila* contain more than one member of the RNase III-like protein family, and the architecture of these proteins often deviates from that of the mammalian prototype.

Human Dicer-1 cleaves precursor miRNAs (pre-miRNAs) and processes longer double strands to siRNAs (Provost et al. 2002; Zhang et al. 2002) in an ATP-independent manner. Recently, the arrangement and function of its domains were investigated biochemically, and a model was derived in which the two RNase III domains form a single pseudo-dimeric active centre (Zhang et al. 2004). The dsRBD of Dicer was shown to have only a minor impact on the in vitro function, while the PAZ domain has an activity in RNA binding and recognition, serving as a molecular ruler for processing of RNA substrates into products of correct size. This is in line with results from structural studies which showed similarities of the PAZ domains of the Ago1 and Ago2 proteins from *D. melanogaster* with the OB fold and a preferential interaction with the 3' protruding ends of siRNAs (Lingel et al. 2003; Ma et al. 2004; Song et al. 2003; Yan et al. 2003; Zhang et al. 2004). While a function of the DUF283 domain is unknown in general, for human Dicer the function of the helicase domain is also not clear, particularly as the enzyme is active in the absence of ATP. However, unwinding activities of RNA helicases have rarely been proven experimentally, and annotated helicases can adopt other functions, like replacing proteins from double-stranded RNA targets, as shown recently (Fairman et al. 2004). Proteins are displaced by helicases in the presence of ATP, and this dependence would also be expected for an analogous function of the helicase domain in the human Dicer. This possibility, however, has not been tested experimentally, yet it is not contradicted by the ATP-independence of the cleavage reaction of the human Dicer-1.

Also *C. elegans* has a single Dicer homologue, DCR-1, that shows an identical domain structure and functions in both RNAi and miRNA pathways (Grishok et al. 2001; Knight and Bass 2001). Finally, this architecture is also found in one of the two *D. melanogaster* Dicers (DCR-2; Fig. 3). Despite this similarity, the enzyme is functionally different, as it is specialised in the processing of long double-stranded RNA, which it cleaves in an ATP-dependent manner (Nykänen et al. 2001). The other *Drosophila* Dicer, Dicer-1, is devoid of the N-terminal helicase domain and functions in miRNA processing (Lee et al. 2004). Such variations both in the domain composition and in function seem to be a recurrent theme among members of the family of RNase III-like proteins. The two *N. crassa* proteins Dcr-1 and Dcr-2 (Fig. 3), for example, both lack the PAZ domain and, additionally, Dcr-1 is devoid of the dsRBD and Dcr-2 of the DUF283 domain. This suggests that these two proteins would also have distinct functions, and it was therefore surprising, that they were found to be redundant in transgene-mediated PTGS (Catalanotto et al. 2004). In *A. thaliana*, on the other hand, four genes for Dicer-like proteins are known, DCL1–DCL4, which all differ in their domain composition (Fig. 3). These proteins also differ in their respective functions, with DCL1 being involved in miRNA, DCL2 in viral siRNA and DCL3 in endogenous siRNA biogenesis (Xie et al. 2004, and references therein). DCL4 localises to the nucleus, and exists in a complex with DRB4, one of four HYL1 homologues (Hiraguri et al. 2005); its cellular function, however, is unknown at present. For the *A. thaliana* Dicers with known function, there is no clearcut correlation between their domain composition and cellular activity. DCL2, for example, as an enzyme processing viral RNA, rather than DCL1 (formerly named CARPEL FACTORY), which is involved in miRNA biogenesis, would be expected to contain two dsRBDs. However, these proteins are not detached units but they functionally (and often physically) associate with other proteins. Thus, the association of DCL1 with HYL1 that contains two dsRBDs might confer the required selectivity and additional binding capacity for viral RNAs. Functional association can also be attributed to DCL3 and the RdRP RDR2, both of which are required for generation of endogenous siRNA targeting heterochromatin formation in *A. thaliana*. It is worth noting that the requirement of RdRPs in RNA-mediated gene regulation is surprisingly common, and they function in many model organisms, including *N. crassa* (Catalanotto et al. 2004), *C. elegans* (Smardon et al. 2000), *D. discoideum* (Martens et al. 2002), and in plants in general (Dalmay et al. 2000), with the exception of mammals and *D. melanogaster*, which lack them.

Presumably the most substantial deviation from the domain architecture of RNase III-like proteins as seen in mammals is that of the Dicers in *D. discoideum* and *T. thermophila* (Fig. 3), which are both made up of only two different protein domains each. DrnA and DrnB, the Dicers of *D. discoideum*, have their dsRBD unusually positioned an the N-terminus. This is unusual when compared with the case for the Dicers of other organisms, but all known dsRBDs in *Dictyostelium* are located N-terminally. Apart from the dsRBDs and the functionally essential two RNase III domains, the *Dictyostelium* Dicers lack all other protein domains of the mammalian Dicer (Martens et al. 2002). It is worth noting that DrnA and DrnB show only 23% sequence identity among each other that is confined to their protein domains, indicative of non-redundant cellular functions. In terms of their domain composition, but not arrangement, these proteins resemble most closely Drosha (Fig. 3), a nuclear enzyme that processes miRNA primary transcripts in mammals and *D. melanogaster* (Lee et al. 2003).

The three Dicers of *T. thermophila*, Dcr1p, Dcr2p and Dcl1p (Fig. 3), have helicase and RNase III domains only. In Dcr1p only one of the two RNase III domains required for activity is well conserved, and a disruption of its gene did not show an obvious phenotype, while Dcr2p was reported to be essential for vegetative growth. The cellular function of Dcl1p, which has the absolute minimum of domains for an RNase III-like enzyme (Fig. 3), is studied best, and the protein was shown to be essential for processing micronuclear transcripts into scan RNAs, which then target DNA elimination in the transcriptionally active macronucleus (Mochizuki and Gorovsky 2005). Details on this RNA-mediated process to eliminate unnecessary DNA elements in ciliates can be found in two excellent recent reviews (Matzke and Birchler 2005; Mochizuki and Gorovsky 2004).

In summary, members of the RNase III-like family have notably divergent protein domains and functions in different organisms, yet all eukaryotic enzymes characterised functionally so far are involved in mechanisms of RNA-mediated gene regulation. Among these, Dcl1-p from *T. thermophila* clearly has the simplest domain structure, and the Dicers from mammals, DCR-1 from *D. melanogaster* and DCL1 and DCL2 from *A. thaliana*, which all have four additional protein domains, have the more complex architecture. This mosaicity of RNase III-like enzymes strongly argues for a diversification in RNA-mediated gene regulation processes in different eukaryotic organisms. It is likely that not only the key players are different in terms of their molecular architecture, which could be compensated by domain shuffling between proteins, which was, for example, proposed for the helicase domain that is present in all RdRPs of *D. discoideum* but that is absent in the Dicers of this organism (Martens et al., 2002), but also that, and particularly with respect to the absence or presence of RdRPs, varying mechanisms of RNAi are likely to have evolved in different organisms. The elucidation of these differences seems at present to be one of the greatest and most fascinating challenges in RNA biochemistry.

4
Cellular Response to Double-Stranded RNA

In the previous section, cellular enzymes were described that all react to the presence of double-stranded RNA in the cell. So, the question is: which of the mechanisms is switched on if a cell is exposed to double-stranded RNA? In general, there are two main determinants and these are the length of the double strand in the respective RNA molecule and its subcellular localisation. However, this discrimination can be leaky, leading to overlaps of different cellular mechanisms, as monitored by unexpected experimental data. Figure 4 summarises major cellular processes that are triggered by double-stranded RNAs.

4.1
Nuclear Processes

Pri-miRNAs show imperfect double strands and are expressed *in trans*, i.e. from a different locus than the mRNA, whose translation the processed miRNA controls in the cytosol (summarised in Ambros, 2004). Pri-miRNAs are processed in the nucleus into pre-miRNAs by a member of the RNase III-

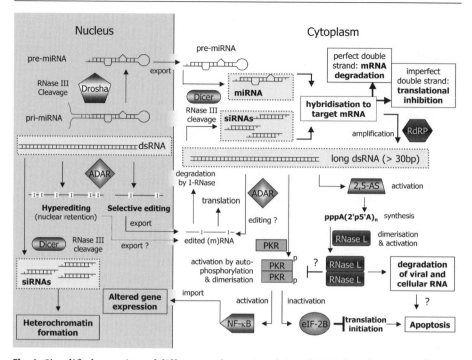

Fig. 4 Simplified overview of different pathways involving dsRNA in eukaryotic cells. The fate of the dsRNA depends on the length of the double strand and the subcellular localisation, as discussed in the text in detail. dsRNAs of different size are *highlighted* and *enclosed with broken lines*. Processes involving dsRNA in the nucleus are shown on the *left* and those in the cytoplasm are shown on the *right*. Note that this overview summarises pathways that are not necessarily expected to be co-current. *pri-miRNA* primary microRNA, *pre-miRNA* precursor microRNA, *miRNA* microRNA, *siRNA* small interfering RNA

like family, in *D. melanogaster* and mammals by Drosha and in other organisms by one of their Dicers. Pre-miRNAs are subsequently exported from the nucleus by exportin-5 (Lee et al. 2003; Yi et al. 2003) and enter the cytoplasmic RNAi pathway. Processing of pri-miRNA into pre-miRNA seems to be specific and governed by the imperfect helical secondary structure of the substrate RNA.

Longer, perfect double-stranded RNAs can have a different fate, as they are also substrates of nuclear ADARs that minimally require helix lengths of about 30 base pairs (Bass and Weintraub 1988). Such RNAs can be present in the nucleus from either antisense transcription, transcriptional read-through, transcription of repetitive DNA elements or viral infection. Hyper-editing of these RNAs leads to RNA molecules with extensive A → I modifications, which associate with a nuclear matrix associated protein complex, resulting in their nuclear retention (Zhang and Carmichael 2001). Hyper-editing is particularly observed for RNAs of viral origin, and presumably the long splice variant of ADAR1, which is expressed from an interferon-inducible promoter, is responsible for this activity (George and Samuel 1999). Editing can also function in preventing the dicing of endogenous sequences in the cytoplasm, as shown for *C. elegans*. ADAR mutants of the nematode are deficient in chemotaxis, and this phenotype could be rescued by additional mutations of components essential for RNAi (Tonkin and Bass 2003). Editing of specific sites, particularly in certain pre-mRNAs like that of GluR-B, also occurs in the nucleus; however, double strands of these substrates are usually not as extensive and imperfect.

Thus, the extent of the perfect double strand largely determines whether an RNA is edited promiscuously or selectively in the nucleus.

While most of the RNAi machinery localises to the cytoplasm, small double-stranded RNA can also affect gene expression and chromatin structure in the nucleus. Such RNA molecules, termed siRNAs, short heterochromatic RNAs or repeat-associated siRNAs, are endogenously present for heterochromatin formation by histone methylation and/or DNA methylation of centromeric repeats and transposons (details of these, among different organisms, highly variable mechanisms are reviewed in Chap. 9 by Paulsen et al. and by Matzke and Birchler 2005). The triggering siRNAs are derived by dicing of longer double-stranded RNAs that can originate from antisense, bidirectional transcription or the action of an RdRP.

RNA-directed DNA methylation was discovered originally in viroid-infected plants (Wassenegger et al. 1994), a finding that was under considerable debate at that time before the advent of RNAi. Later on, siRNAs (21–24 nucleotides), derived from long double-stranded RNA, were shown to be required for this process (Aufsatz et al. 2002). In *Schizosaccharomyces pombe*, the association of an RdRP with centromeric repeats was shown by chromatin immunoprecipitation, suggesting a function in the generation of double-stranded RNAs (Volpe et al. 2003). Nuclear dicing of these molecules would then result in siRNAs corresponding to centromeric DNA sequences, which had been cloned earlier (Reinhart and Bartel 2002). Fission yeast, however, does not show DNA methylation, and siRNAs trigger heterochromatin formation by directing methylation of histones (Volpe et al. 2002). In mammals, siRNAs have been suggested to induce DNA methylation in cell culture experiments (Morris et al. 2004). It remains to be determined whether nuclear siRNAs can originate from the cytoplasm, as has been discussed for plants (Matzke and Birchler 2005).

Summarising, long double strands in the nucleus can trigger both mechanisms, RNA editing and RNA-mediated heterochromatin formation. How does the cell discriminate between the two possible pathways? At present, an answer to this question is difficult, since only limited information is available on this issue. However, presumably the cell can discriminate between the two pathways. It is unlikely that the required information resides within the double-stranded RNA molecules themselves, as sequence information cannot be read from perfect A-type helices, as laid out earlier. Rather, it would seem possible that local and global nuclear concentrations of respective RNA molecules play a role, as well as their protein decoration. Proteins might thus define the origin of the double-stranded RNA, allowing the cell to start the appropriate pathway. However, at present, it cannot be excluded that there is indeed an overlap, or at least that the discrimination of pathways is leaky, as has been discussed (Wang and Carmichael 2004).

4.2
Cytoplasmic Processes

Both pre-miRNAs and long double-stranded RNAs are cleaved in the cytosol by enzymes of the Dicer family into short duplexes of about 21 nucleotides, resulting in miRNAs and siRNAs, respectively. Dicers presumably recognise specifically the secondary structure of pre-miRNAs, but for long double-stranded RNAs, they have to compete with other cytoplasmic enzymes, predominantly PKR and 2,5-AS.

One of the two strands of the siRNAs produced by Dicers can hybridise by base complementarity to a target mRNA in a complex, protein-assisted way. The position and perfectness of the newly formed stretch of double-stranded RNA largely contributes to the fate of the target mRNA: perfect double strands are endonucleolytically degraded, while imperfect double strands result particularly when located towards the $3'$ end of the mRNA, in translational repression (Meister and Tuschl 2004, and references therein). The potential of miRNAs and siRNAs to act both as a translational repressor and in mRNA degradation was concluded from the observations that miRNA can act as siRNAs and vice versa (Doench et al. 2003; Hutvagner and Zamore 2002; Saxena et al. 2003). However, it is not clear at present, whether this potential is frequently realised in vivo. In organisms endowed with RdRPs, the duplex formed between the target mRNA and the single-stranded small RNA might additionally, or alternatively, serve as a substrate for the enzymes, resulting in (re)amplification of the long double strand.

Next to these processes targeting specific mRNAs, eukaryotic cells have developed several defence mechanisms against long double-stranded RNA, particularly of viral origin, that lead to unspecific and considerably more dramatic changes in the cell.

These include, among others, the PKR pathway in mammals, in which binding of the RNA double strand by PKR leads to its activation by autophosphorylation and dimerisation. Subsequently, several target proteins are phosphorylated. This leads, among others, to an activated transcription factor NF-κB that alters gene expression, or via phosphorylation of eIF-2α to an inactivated eIF-2B, which ultimately results in apoptosis by breakdown of translation initiation. Also, binding of the 2,5-AS to double-stranded RNA results in the activation of RNase L, a process mediated by the presence of $2'-5'$ connected polyA molecules. RNase L then degrades cellular and viral RNA in an unspecific manner, and thus can also lead to apoptosis (summarised in Stark et al. 1998). A function of the cytoplasmic ADAR in antiviral defence has also been discussed; however, details of this are unknown as yet (Wong et al. 2003).

Both PKR and 2,5-AS require for their activation RNA double strands larger than 30 and 40 base pairs, respectively (Manche et al. 1992; Minks et al. 1979). This clearly hampered the use of long double strands in RNAi experiments in mammals, and restricted this design to cells devoid of a PKR response, like undifferentiated embryonic cells (Billy et al. 2001). To circumvent the largely unspecific effects triggered by long double-stranded RNA, the direct use of siRNA in cultured cells was introduced, which greatly advanced research of RNAi in mammalian cells (Elbashir et al. 2001). There is convincing evidence that these siRNAs can also trigger unspecific effects, some of which could be attributed to the PKR and RNase L pathways (Persengiev et al. 2004; Scacheri et al. 2004; Sledz et al. 2003). In view of the RNA substrate requirement of these processes, these observations were surprising; however, they are not made in general. Rather, they might be explained by the preferential binding of PKR to certain sequences (Circle et al. 1997; Spanggord et al. 2002) which might have been present fortuitously in the siRNAs used in those experiments. Since siRNA triggering of these effects is concentration-dependent, PKR might have bound several siRNAs, possibly even ligated siRNAs, as they were observed in another context (Martinez et al. 2002).

Summarising, small double-stranded RNAs with perfect or imperfect base complementarity trigger as siRNAs or miRNAs in the cytoplasm regulatory processes at the post-transcriptional and translational levels, respectively. In exceptional cases, these small RNA duplexes also can cause in mammals unspecific effects that are mediated by the PKR and 2,5-AS pathways. Normally, these are evoked by considerably longer RNA double strands, and, in general, they are confined to mammalian organisms. Other organisms or mammalian cells that are devoid of the PKR response can also use long RNA double strands to elicit a specific regulatory response via base complementarity to a target sequence.

References

Ambros V (2004) The functions of animal microRNAs. Nature 431:350–355

Aravind L, Watanabe H, Lipman DJ, Koonin EV (2000) Lineage-specific loss and divergence of functionally linked genes in eukaryotes. Proc Natl Acad Sci USA 97:11319–11324

Aufsatz W, Mette MF, van der Winden J, Matzke AJ, Matzke M (2002) RNA-directed DNA methylation in Arabidopsis. Proc Natl Acad Sci USA 99 Suppl 4:16499–16506

Ban N, Nissen P, Hansen J, Moore PB, Steitz TA (2000) The complete atomic structure of the large ribosomal subunit at 2.4 Åresolution. Science 289:905–920

Bartsch H, Voigtsberger S, Baumann G, Morano I, Luther HP (2004) Detection of a novel sense-antisense RNA-hybrid structure by RACE experiments on endogenous Troponin I antisense RNA. RNA 10:1215–1224

Bass BL, Weintraub H (1988) An unwinding activity that covalently modifies its double-stranded RNA substrate. Cell 55:1089–1098

Bass BL, Hurst SR, Singer JD (1994) Binding properties of newly identified Xenopus proteins containing dsRNA-binding motifs. Curr Biol 4: 301–314

Beal PA (2005) Duplex RNA-binding enzymes: headliners from neurobiology, virology, and development. ChemBioChem 6:257–266

Billy E, Brondani V, Zhang H, Muller U, Filipowicz W (2001) Specific interference with gene expression induced by long, double-stranded RNA in mouse embryonal teratocarcinoma cell lines. Proc Natl Acad Sci USA 98:14428–14433

Brown BA II, Lowenhaupt K, Wilbert CM, Hanlon EB, Rich A (2000) The Zα domain of the editing enzyme dsRNA adenosine deaminase binds left-handed Z-RNA as well as Z-DNA. Proc Natl Acad Sci USA 97:13532–13536

Bycroft M, Grunert S, Murzin AG, Proctor M, St Johnston D (1995) NMR solution structure of a dsRNA binding domain from Drosophila Staufen protein reveals homology to the N-terminal domain of ribosomal protein S5. EMBO J 14:3563–3571

Carmell MA, Xuan Z, Zhang MQ, Hannon GJ (2002) The Argonaute family: tentacles that reach into RNAi, developmental control, stem cell maintenance, and tumorigenesis. Genes Dev 16:2733–2742

Catalanotto C, Pallotta M, ReFalo P, Sachs MS, Vayssie L, Macino G, Cogoni C (2004) Redundancy of the two dicer genes in transgene-induced posttranscriptional gene silencing in Neurospora crassa. Mol Cell Biol 24:2536–2545

Cerutti L, Mian N, Bateman A (2000) Domains in gene silencing and cell differentiation proteins: the novel PAZ domain and redefinition of the Piwi domain. Trends Biochem Sci 25:481–482

Chen J, Sun M, Kent WJ, Huang X, Xie H, Wang W, Zhou G, Shi RZ, Rowley JD (2004) Over 20% of human transcripts might form sense-antisense pairs. Nucleic Acids Res 32:4812–4820

Chicas A, Cogoni C, Macino G (2004) RNAi-dependent and RNAi-independent mechanisms contribute to the silencing of RIPed sequences in Neurospora crassa. Nucleic Acids Res 32:4237–4243

Circle DA, Neel OD, Robertson HD, Clarke PA, Mathews MB (1997) Surprising specificity of PKR binding to delta agent genomic RNA. RNA 3:438–448

Conte MR, Conn GL, Brown T, Lane AN (1997) Conformational properties and thermodynamics of the RNA duplex r(CGCAAAUUUGCG)2: comparison with the DNA analogue d(CGCAAATTTGCG)2. Nucleic Acids Res 25:2627–2634

Dahary D, Elroy-Stein O, Sorek R (2005) Naturally occurring antisense: transcriptional leakage or real overlap? Genome Res 15:364–368

Dalmay T, Hamilton A, Rudd S, Angell S, Baulcombe DC (2000) An RNA-dependent RNA polymerase gene in Arabidopsis is required for posttranscriptional gene silencing mediated by a transgene but not by a virus. Cell 101:543–553

DeLano WL (2002) The PyMOL molecular graphics system. http://www.pymol.org

Doench JG, Petersen CP, Sharp PA (2003) siRNAs can function as miRNAs. Genes Dev 17:438–442

Draper DE (1999) Themes in RNA-protein recognition. J Mol Biol 293:255–270

Drew HR, Wing RM, Takano T, Broka C, Tanaka S, Itakura K, Dickerson RE (1981) Structure of a B-DNA dodecamer: conformation and dynamics. Proc Natl Acad Sci USA 78:2179–2183

Eckmann CR, Jantsch MF (1997) Xlrbpa, a double-stranded RNA-binding protein associated with ribosomes and heterogeneous nuclear RNPs. J Cell Biol 138:239–253

Elbashir SM, Harborth J, Lendeckel W, Yalcin A, Weber K, Tuschl T (2001) Duplexes of 21-nucleotide RNAs mediate RNA interference in cultured mammalian cells. Nature 411:494–498

Fairman ME, Maroney PA, Wang W, Bowers HA, Gollnick P, Nilsen TW, Jankowsky E (2004) Protein displacement by DExH/D "RNA helicases" without duplex unwinding. Science 304:730–734

Farrell PJ, Balkow K, Hunt T, Jackson RJ, Trachsel H (1977) Phosphorylation of initiation factor eIF-2 and the control of reticulocyte protein synthesis. Cell 11:187–200

Fire A, Xu S, Montgomery MK, Kostas SA, Driver SE, Mello CC (1998) Potent and specific genetic interference by double-stranded RNA in Caenorhabditis elegans. Nature 391:806–811

Galabru J, Hovanessian A (1987) Autophosphorylation of the protein kinase dependent on double-stranded RNA. J Biol Chem 262:15538–15544

Gatignol A, Buckler-White A, Berkhout B, Jeang KT (1991) Characterization of a human TAR RNA-binding protein that activates the HIV-1 LTR. Science 251:1597–1600

Gazzani S, Lawrenson T, Woodward C, Headon D, Sablowski R (2004) A link between mRNA turnover and RNA interference in Arabidopsis. Science 306: 1046–1048

George CX, Samuel CE (1999) Human RNA-specific adenosine deaminase ADAR1 transcripts possess alternative exon 1 structures that initiate from different promoters, one constitutively active and the other interferon inducible. Proc Natl Acad Sci USA 96:4621–4626

Gräf S, Borisova BE, Nellen W, Steger G, Hammann C (2004) A database search for double-strand containing RNAs in Dictyostelium discoideum. Biol Chem 385:961–965

Grishok A, Pasquinelli AE, Conte D, Li N, Parrish S, Ha I, Baillie DL, Fire A, Ruvkun G, Mello CC (2001) Genes and mechanisms related to RNA interference regulate expression of the small temporal RNAs that control C. elegans developmental timing. Cell 106:23–34

Hall K, Cruz P, Tinoco I, Jr., Jovin TM, van de Sande JH (1984) "Z-RNA"–a left-handed RNA double helix. Nature 311:584–586

Hall KB (2002) RNA-protein interactions. Curr Opin Struct Biol 12:283–288

Han MH, Goud S, Song L, Fedoroff N (2004) The Arabidopsis double-stranded RNA-binding protein HYL1 plays a role in microRNA-mediated gene regulation. Proc Natl Acad Sci USA 101:1093–1098

Higuchi M, Single FN, Kohler M, Sommer B, Sprengel R, Seeburg PH (1993) RNA editing of AMPA receptor subunit GluR-B: a base-paired intron-exon structure determines position and efficiency. Cell 75:1361–1370

Higuchi M, Maas S, Single FN, Hartner J, Rozov A, Burnashev N, Feldmeyer D, Sprengel R, Seeburg PH (2000) Point mutation in an AMPA receptor gene rescues lethality in mice deficient in the RNA-editing enzyme ADAR2. Nature 406:78–81

Hildebrandt M, Nellen W (1992) Differential antisense transcription from the Dictyostelium EB4 gene locus: implications on antisense-mediated regulation of mRNA stability. Cell 69:197–204

Hiraguri A, Itoh R, Kondo N, Nomura Y, Aizawa D, Murai Y, Koiwa H, Seki M, Shinozaki K, Fukuhara T (2005) Specific interactions between Dicer-like proteins and HYL1/DRB-family dsRNA-binding proteins in Arabidopsis thaliana. Plant Mol Biol 57:173–188

Humphrey W, Dalke A, Schulten K (1996) VMD: visual molecular dynamics. J Mol Graph 14:33–38, 27–28

Hutvagner G, Zamore PD (2002) A microRNA in a multiple-turnover RNAi enzyme complex. Science 297:2056–2060

Kharrat A, Macias MJ, Gibson TJ, Nilges M, Pastore A (1995) Structure of the dsRNA binding domain of E. coli RNase III. EMBO J 14:3572–3584

Klaue Y, Kallman AM, Bonin M, Nellen W, Ohman M (2003) Biochemical analysis and scanning force microscopy reveal productive and nonproductive ADAR2 binding to RNA substrates. RNA 9:839–846

Knight SW, Bass BL (2001) A role for the RNase III enzyme DCR-1 in RNA interference and germ line development in Caenorhabditis elegans. Science 293:2269–2271

Kolb FA, Malmgren C, Westhof E, Ehresmann C, Ehresmann B, Wagner EG, Romby P (2000) An unusual structure formed by antisense-target RNA binding involves an extended kissing complex with a four-way junction and a side-by-side helical alignment. RNA 6:311–324

Lavorgna G, Dahary D, Lehner B, Sorek R, Sanderson CM, Casari G (2004) In search of antisense. Trends Biochem Sci 29:88–94

Lee K, Fajardo MA, Braun RE (1996) A testis cytoplasmic RNA-binding protein that has the properties of a translational repressor. Mol Cell Biol 16:3023–3034

Lee Y, Ahn C, Han J, Choi H, Kim J, Yim J, Lee J, Provost P, Radmark O, Kim S, Kim VN (2003) The nuclear RNase III Drosha initiates microRNA processing. Nature 425:415–419

Lee YS, Nakahara K, Pham JW, Kim K, He Z, Sontheimer EJ, Carthew RW (2004) Distinct roles for Drosophila Dicer-1 and Dicer-2 in the siRNA/miRNA silencing pathways. Cell 117:69–81

Lehner B, Williams G, Campbell RD, Sanderson CM (2002) Antisense transcripts in the human genome. Trends Genet 18:63–65

Leontis NB, Westhof E (2001) Geometric nomenclature and classification of RNA base pairs. RNA 7:499–512

Lingel A, Simon B, Izaurralde E, Sattler M (2003) Structure and nucleic-acid binding of the Drosophila Argonaute 2 PAZ domain. Nature 426: 465–469

Liu Q, Rand TA, Kalidas S, Du F, Kim HE, Smith DP, Wang X (2003) R2D2, a bridge between the initiation and effector steps of the Drosophila RNAi pathway. Science 301:1921–1925

Liu Y, Lei M, Samuel CE (2000) Chimeric double-stranded RNA-specific adenosine deaminase ADAR1 proteins reveal functional selectivity of double-stranded RNA-binding domains from ADAR1 and protein kinase PKR. Proc Natl Acad Sci USA 97:12541–12546

Ma JB, Ye K, Patel DJ (2004) Structural basis for overhang-specific small interfering RNA recognition by the PAZ domain. Nature 429:318–322

Manche L, Green SR, Schmedt C, Mathews MB (1992) Interactions between double-stranded RNA regulators and the protein kinase DAI. Mol Cell Biol 12:5238–5248

Martens H, Novotny J, Oberstrass J, Steck TL, Postlethwait P, Nellen W (2002) RNAi in Dictyostelium: the role of RNA-directed RNA polymerases and double-stranded RNase. Mol Biol Cell 13:445–453

Martinez J, Patkaniowska A, Urlaub H, Luhrmann R, Tuschl T (2002) Single-stranded antisense siRNAs guide target RNA cleavage in RNAi. Cell 110: 563–574

Matzke MA, Birchler JA (2005) RNAi-mediated pathways in the nucleus. Nat Rev Genet 6:24–35

Meister G, Tuschl T (2004) Mechanisms of gene silencing by double-stranded RNA. Nature 431:343–349

Melcher T, Maas S, Herb A, Sprengel R, Seeburg PH, Higuchi M (1996) A mammalian RNA editing enzyme. Nature 379:460–464

Messias AC, Sattler M (2004) Structural basis of single-stranded RNA recognition. Acc Chem Res 37:279–287

Micklem DR, Adams J, Grunert S, St Johnston D (2000) Distinct roles of two conserved Staufen domains in oskar mRNA localization and translation. EMBO J 19:1366–1377

Minks MA, West DK, Benvin S, Baglioni C (1979) Structural requirements of double-stranded RNA for the activation of 2′,5′-oligo(A) polymerase and protein kinase of interferon-treated HeLa cells. J Biol Chem 254:10180–10183

Misra S, Crosby MA, Mungall CJ, Matthews BB, Campbell KS, Hradecky P, Huang Y, Kaminker JS, Millburn GH, Prochnik SE, Smith CD, Tupy JL, Whitfied EJ, Bayraktaroglu L, Berman BP, Bettencourt BR, Celniker SE, de Grey AD, Drysdale RA, Harris NL, Richter J, Russo S, Schroeder AJ, Shu SQ, Stapleton M, Yamada C, Ashburner M, Gelbart WM, Rubin GM, Lewis SE (2002) Annotation of the Drosophila melanogaster euchromatic genome: a systematic review. Genome Biol 3:research0083.1–0083.22

Mochizuki K, Gorovsky MA (2004) Small RNAs in genome rearrangement in Tetrahymena. Curr Opin Genet Dev 14:181–187

Mochizuki K, Gorovsky MA (2005) A Dicer-like protein in Tetrahymena has distinct functions in genome rearrangement, chromosome segregation, and meiotic prophase. Genes Dev 19:77–89

Morris KV, Chan SW, Jacobsen SE, Looney DJ (2004) Small interfering RNA-induced transcriptional gene silencing in human cells. Science 305:1289–1292

Munroe SH (2004) Diversity of antisense regulation in eukaryotes: multiple mechanisms, emerging patterns. J Cell Biochem 93:664–671

Nanduri S, Carpick BW, Yang Y, Williams BR, Qin J (1998) Structure of the double-stranded RNA-binding domain of the protein kinase PKR reveals the molecular basis of its dsRNA-mediated activation. EMBO J 17: 5458–5465

Nykänen A, Haley B, Zamore PD (2001) ATP requirements and small interfering RNA structure in the RNA interference pathway. Cell 107:309–321

Park H, Davies MV, Langland JO, Chang HW, Nam YS, Tartaglia J, Paoletti E, Jacobs BL, Kaufman RJ, Venkatesan S (1994) TAR RNA-binding protein is an inhibitor of the interferon-induced protein kinase PKR. Proc Natl Acad Sci USA 91:4713–4717

Patel RC, Sen GC (1998) PACT, a protein activator of the interferon-induced protein kinase, PKR. EMBO J 17:4379–4390

Patterson JB, Samuel CE (1995) Expression and regulation by interferon of a double-stranded-RNA-specific adenosine deaminase from human cells: evidence for two forms of the deaminase. Mol Cell Biol 15:5376–5388

Persengiev SP, Zhu X, Green MR (2004) Nonspecific, concentration-dependent stimulation and repression of mammalian gene expression by small interfering RNAs (siRNAs). RNA 10:12–18

Peters GA, Hartmann R, Qin J, Sen GC (2001) Modular structure of PACT: distinct domains for binding and activating PKR. Mol Cell Biol 21:1908–1920

Polson AG, Bass BL, Casey JL (1996) RNA editing of hepatitis delta virus antigenome by dsRNA-adenosine deaminase. Nature 380:454–456

Polson AG, Crain PF, Pomerantz SC, McCloskey JA, Bass BL (1991) The mechanism of adenosine to inosine conversion by the double-stranded RNA unwinding/modifying activity: a high-performance liquid chromatography–mass spectrometry analysis. Biochemistry 30:11507–11514

Provost P, Dishart D, Doucet J, Frendewey D, Samuelsson B, Radmark O (2002) Ribonuclease activity and RNA binding of recombinant human Dicer. EMBO J 21:5864–5874

Ramos A, Grunert S, Adams J, Micklem DR, Proctor MR, Freund S, Bycroft M, St Johnston D, Varani G (2000) RNA recognition by a Staufen double-stranded RNA-binding domain. EMBO J 19:997–1009

Reinhart BJ, Bartel DP (2002) Small RNAs correspond to centromere heterochromatic repeats. Science 297:1831

Rich A (2004) The excitement of discovery. Annu Rev Biochem 73: 1–37

Røsok Ø, Sioud M (2004) Systematic identification of sense–antisense transcripts in mammalian cells. Nat Biotechnol 22:104–108

Rothenburg S, Deigendesch N, Dittmar K, Koch-Nolte F, Haag F, Lowenhaupt K, Rich A (2005) A PKR-like eukaryotic initiation factor 2alpha kinase from zebrafish contains Z-DNA binding domains instead of dsRNA binding domains. Proc Natl Acad Sci USA 102:1602–1607

Ryter JM, Schultz SC (1998) Molecular basis of double-stranded RNA-protein interactions: structure of a dsRNA-binding domain complexed with dsRNA. EMBO J 17:7505–7513

Saunders LR, Barber GN (2003) The dsRNA binding protein family: critical roles, diverse cellular functions. FASEB J 17:961–983

Saxena S, Jonsson ZO, Dutta A (2003) Small RNAs with imperfect match to endogenous mRNA repress translation. Implications for off-target activity of small inhibitory RNA in mammalian cells. J Biol Chem 278:44312–44319

Scacheri PC, Rozenblatt-Rosen O, Caplen NJ, Wolfsberg TG, Umayam L, Lee JC, Hughes CM, Shanmugam KS, Bhattacharjee A, Meyerson M, Collins FS (2004) Short interfering RNAs can induce unexpected and divergent changes in the levels of untargeted proteins in mammalian cells. Proc Natl Acad Sci USA 101:1892–1897

Scadden AD, Smith CW (2001) Specific cleavage of hyper-edited dsRNAs. EMBO J 20:4243–4252

Schluenzen F, Tocilj A, Zarivach R, Harms J, Gluehmann M, Janell D, Bashan A, Bartels H, Agmon I, Franceschi F, Yonath A (2000) Structure of functionally activated small ribosomal subunit at 3.3 angstroms resolution. Cell 102:615–623

Schroeder R, Barta A, Semrad K (2004) Strategies for RNA folding and assembly. Nat Rev Mol Cell Biol 5:908–919

Schwartz T, Rould MA, Lowenhaupt K, Herbert A, Rich A (1999) Crystal structure of the Z domain of the human editing enzyme ADAR1 bound to left-handed Z-DNA. Science 284:1841–1845

Sledz CA, Holko M, de Veer MJ, Silverman RH, Williams BR (2003) Activation of the interferon system by short-interfering RNAs. Nat Cell Biol 5:834–839

Smardon A, Spoerke JM, Stacey SC, Klein ME, Mackin N, Maine EM (2000) EGO-1 is related to RNA-directed RNA polymerase and functions in germ-line development and RNA interference in C. elegans. Curr Biol 10:169–178

Song JJ, Liu J, Tolia NH, Schneiderman J, Smith SK, Martienssen RA, Hannon GJ, Joshua-Tor L (2003) The crystal structure of the Argonaute2 PAZ domain reveals an RNA binding motif in RNAi effector complexes. Nat Struct Biol 10:1026–1032

Spanggord RJ, Vuyisich M, Beal PA (2002) Identification of binding sites for both dsRBMs of PKR on kinase-activating and kinase-inhibiting RNA ligands. Biochemistry 41:4511–4520

St Johnston D, Beuchle D, Nusslein-Volhard C (1991) Staufen, a gene required to localize maternal RNAs in the Drosophila egg. Cell 66:51–63

St Johnston D, Brown NH, Gall JG, Jantsch M (1992) A conserved double-stranded RNA-binding domain. Proc Natl Acad Sci USA 89:10979–10983

Stark GR, Kerr IM, Williams BR, Silverman RH, Schreiber RD (1998) How cells respond to interferons. Annu Rev Biochem 67:227–264

Stefl R, Skrisovska L, Allain FH (2005) RNA sequence- and shape-dependent recognition by proteins in the ribonucleoprotein particle. EMBO Rep 6:33–38

Stein P, Svoboda P, Anger M, Schultz RM (2003) RNAi: mammalian oocytes do it without RNA-dependent RNA polymerase. RNA 9:187–192

Stephens OM, Haudenschild BL, Beal PA (2004) The binding selectivity of ADAR2's dsRBMs contributes to RNA-editing selectivity. Chem Biol 11:1239–1250

Stolc V, Samanta MP, Tongprasit W, Sethi H, Liang S, Nelson DC, Hegeman A, Nelson C, Rancour D, Bednarek S, Ulrich EL, Zhao Q, Wrobel RL, Newman CS, Fox BG, Phillips GN, Jr., Markley JL, Sussman MR (2005) Identification of transcribed sequences in Arabidopsis thaliana by using high-resolution genome tiling arrays. Proc Natl Acad Sci USA 102:4453–4458

Tabara H, Yigit E, Siomi H, Mello CC (2002) The dsRNA binding protein RDE-4 interacts with RDE-1, DCR-1, and a DExH-box helicase to direct RNAi in C. elegans. Cell 109:861–871

Tian B, Bevilacqua PC, Diegelman-Parente A, Mathews MB (2004) The double-stranded-RNA-binding motif: interference and much more. Nat Rev Mol Cell Biol 5:1013–1023

Tian B, Mathews MB (2001) Functional characterization of and cooperation between the double-stranded RNA-binding motifs of the protein kinase PKR. J Biol Chem 276:9936–9944

Tonkin LA, Bass BL (2003) Mutations in RNAi rescue aberrant chemotaxis of ADAR mutants. Science 302:1725

van Dijk AA, Makeyev EV, Bamford DH (2004) Initiation of viral RNA-dependent RNA polymerization. J Gen Virol 85:1077–1093

Varani G, Nagai K (1998) RNA recognition by RNP proteins during RNA processing. Annu Rev Biophys Biomol Struct 27:407–445

Vazquez F, Gasciolli V, Crete P, Vaucheret H (2004) The nuclear dsRNA binding protein HYL1 is required for microRNA accumulation and plant development, but not posttranscriptional transgene silencing. Curr Biol 14:346–351

Volpe TA, Kidner C, Hall IM, Teng G, Grewal SI, Martienssen RA (2002) Regulation of heterochromatic silencing and histone H3 lysine-9 methylation by RNAi. Science 297:1833–1837

Volpe T, Schramke V, Hamilton GL, White SA, Teng G, Martienssen RA, Allshire RC (2003) RNA interference is required for normal centromere function in fission yeast. Chromosome Res 11:137–146

Wang AH-J, Quigley GJ, Kolpak FJ, Crawford JL, van Boom JH, van der Marel G, Rich A (1979) Molecular structure of a left-handed double-helical DNA fragment at atomic resolution. Nature 282:680–686

Wang Q, Carmichael GG (2004) Effects of length and location on the cellular response to double-stranded RNA. Microbiol Mol Biol Rev 68:432–452

Wimberly BT, Brodersen DE, Clemons WM Jr, Morgan-Warren RJ, Carter AP, Vonrhein C, Hartsch T, Ramakrishnan V (2000) Structure of the 30S ribosomal subunit. Nature 407:327–339

Wong SK, Sato S, Lazinski DW (2003) Elevated activity of the large form of ADAR1 in vivo: very efficient RNA editing occurs in the cytoplasm. RNA 9:586–598

Wu H, Henras A, Chanfreau G, Feigon J (2004) Structural basis for recognition of the AGNN tetraloop RNA fold by the double-stranded RNA-binding domain of Rnt1p RNase III. Proc Natl Acad Sci USA 101:8307–8312

Xie Z, Johansen LK, Gustafson AM, Kasschau KD, Lellis AD, Zilberman D, Jacobsen SE, Carrington JC (2004) Genetic and functional diversification of small RNA pathways in plants. PLoS Biol 2:E104

Yamada K, Lim J, Dale JM, Chen H, Shinn P, Palm CJ, Southwick AM, Wu HC, Kim C, Nguyen M, Pham P, Cheuk R, Karlin-Newmann G, Liu SX, Lam B, Sakano H, Wu T, Yu G, Miranda M, Quach HL, Tripp M, Chang CH, Lee JM, Toriumi M, Chan MM, Tang CC, Onodera CS, Deng JM, Akiyama K, Ansari Y, Arakawa T, Banh J, Banno F, Bowser L, Brooks S, Carninci P, Chao Q, Choy N, Enju A, Goldsmith AD, Gurjal M, Hansen NF, Hayashizaki Y, Johnson-Hopson C, Hsuan VW, Iida K, Karnes M, Khan S, Koesema E, Ishida J, Jiang PX, Jones T, Kawai J, Kamiya A, Meyers C, Nakajima M, Narusaka M, Seki M, Sakurai T, Satou M, Tamse R, Vaysberg M, Wallender EK, Wong C, Yamamura Y, Yuan S, Shinozaki K, Davis RW, Theologis A, Ecker JR (2003) Empirical analysis of transcriptional activity in the Arabidopsis genome. Science 302:842–846

Yan KS, Yan S, Farooq A, Han A, Zeng L, Zhou MM (2003) Structure and conserved RNA binding of the PAZ domain. Nature 426:468–474

Yelin R, Dahary D, Sorek R, Levanon EY, Goldstein O, Shoshan A, Diber A, Biton S, Tamir Y, Khosravi R, Nemzer S, Pinner E, Walach S, Bernstein J, Savitsky K, Rotman G (2003) Widespread occurrence of antisense transcription in the human genome. Nat Biotechnol 21:379–386

Yi R, Qin Y, Macara IG, Cullen BR (2003) Exportin-5 mediates the nuclear export of pre-microRNAs and short hairpin RNAs. Genes Dev 17:3011–3016

Zarling DA, Calhoun CJ, Hardin CC, Zarling AH (1987) Cytoplasmic Z-RNA. Proc Natl Acad Sci USA 84:6117–6121

Zhang H, Kolb FA, Brondani V, Billy E, Filipowicz W (2002) Human Dicer preferentially cleaves dsRNAs at their termini without a requirement for ATP. EMBO J 21:5875–5885

Zhang H, Kolb FA, Jaskiewicz L, Westhof E, Filipowicz W (2004) Single processing center models for human Dicer and bacterial RNase III. Cell 118:57–68

Zhang Z, Carmichael GG (2001) The fate of dsRNA in the nucleus: a p54(nrb)-containing complex mediates the nuclear retention of promiscuously A-to-I edited RNAs. Cell 106:465–475

Nucleic Acids and Molecular Biology, Vol. 17
Wolfgang Nellen, Christian Hammann (Eds.)
Small RNAs
© Springer-Verlag Berlin Heidelberg 2005

Transitive and Systemic RNA Silencing: Both Involving an RNA Amplification Mechanism?

Annick Bleys · Helena van Houdt · Anna Depicker (✉)

Department of Plant Systems Biology, Flanders Interuniversity Institute
for Biotechnology, Ghent University, Technologiepark 927, 9052 Gent, Belgium
annick.bleys@psb.ugent.be, helena.vanhoudt@psb.ugent.be, ann.depicker@psb.ugent.be

Abstract RNA silencing is a conserved regulatory mechanism that plays an important role in genome integrity and defense in eukaryotic organisms. A key molecule in this sequence-specific RNA degradation mechanism is double-stranded RNA, which is processed by an RNase-III like enzyme (Dicer) into small interfering RNAs (siRNAs). The initial pool of siRNAs can be amplified through the action of RNA-dependent RNA polymerases, which could account for the observed spreading of RNA silencing along the target gene (transitive silencing) and throughout the organism (systemic silencing). In this chapter we discuss the mechanism of RNA amplification and its possible involvement in transitive and systemic RNA silencing in different organisms.

1
Introduction

Co-suppression, post-transcriptional gene silencing (PTGS) or RNA silencing in plants can be triggered upon introduction of transgenic DNA. The intentional or unintentional production of double-stranded RNA (dsRNA) from the transgene(s) can induce silencing of both the transgene(s) and the homologous endogenous gene. This type of highly conserved eukaryotic silencing mechanism is designated as RNA interference (RNAi) in animal systems. In general, RNA silencing or RNAi is a two-step process. In the first step RNAi-triggering dsRNA molecules are processed into small duplex molecules, the small interfering RNAs (siRNAs; Zamore et al. 2000). This step is catalyzed by an RNase-III-like enzyme Dicer (Bernstein et al. 2001). The second step, the sequence-specific cleavage of cognate target RNAs, is brought about by siRNAs (Elbashir et al. 2001a) bound to specific proteins, which together form a multicomponent nuclease, the RNA-induced silencing complex (RISC; Hammond et al. 2000). The Dicer/RISC pathway accounts for initiation of the silencing process and the homology-dependent selection and degradation of cognate target RNAs. However, a more complex, branched silencing pathway (Fig. 1) must be envisioned to explain the observations of the catalytic nature of RNAi, especially in *Caenorhabditis elegans*, the production of secondary siRNAs, and spreading of silencing throughout organisms.

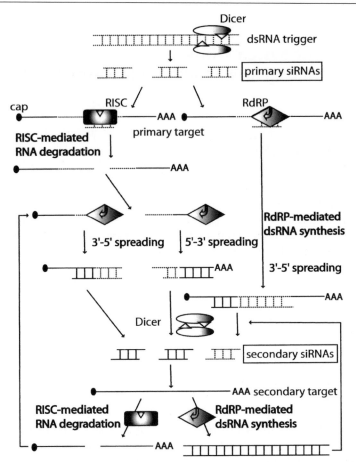

Fig. 1 Illustration of RNA amplification and transitive silencing. The *dotted line* in the primary target represents the sequence homology with the silencing inducer; the *full line* indicates homology with the secondary target. The double-stranded RNA (*dsRNA*) trigger is cleaved by Dicer into primary small interfering RNAs (*siRNAs*), which can subsequently follow two pathways. In the RNA-induced silencing complex (*RISC*)-mediated degradation pathway, they are incorporated into the RISC, where they function as guides for cleavage of homologous RNAs. In the RNA-dependent RNA polymerase (*RdRP*)/Dicer pathway, target transcripts are indirectly degraded. First they serve as templates for RdRP-mediated dsRNA synthesis, followed by cleavage by Dicer into secondary siRNAs. When the primary siRNAs are used as primers for the dsRNA synthesis, 3′–5′ spreading of RNA targeting can occur. In plants however, 5′–3′ spreading is also observed, which could be the result of unprimed dsRNA synthesis starting from the 3′ end of an aberrant RISC-cleaved mRNA, either marked by a hybridizing siRNA or not. Unprimed synthesis could also account for 3′–5′ spreading when the capped 5′ part of the RISC-cleaved messenger RNA is used as a template. The secondary siRNAs produced can again follow the two pathways, resulting in silencing of a secondary target that has no sequence homology to the dsRNA inducer

2
RNA-Dependent RNA Polymerases and RNA Amplification

Early studies of RNAi in *C. elegans* showed the remarkable potency of dsRNA: very small amounts of dsRNA are able to trigger silencing of a vast excess of target messenger RNA (mRNA) throughout the organism (Fire et al. 1998; Sijen et al. 2001). This striking observation suggested the need for an amplification step, without which siRNAs fail to reach sufficient concentrations to accomplish target mRNA degradation (Sijen et al. 2001). Some degree of amplification is obtained by the conversion of the trigger dsRNA into many si-RNAs, but this is not sufficient to bring about such continuous mRNA degradation. RNA-dependent RNA polymerases (RdRPs) are very good candidate enzymes to be involved in the amplification of RNA silencing signals for efficient RNA silencing. Schiebel et al. (1993a, b) showed that the purified tomato RdRP catalyzes the synthesis of dsRNA by using single-stranded RNA (ssRNA) as a template, both in a primed and unprimed fashion. Five years later they successfully cloned the gene (Schiebel et al. 1998). Putative homologues of this gene are found in plants, fungi, and worms. Mutations in the *qde-1* gene of *Neurospora crassa* (Cogoni and Macino 1999), the *sgs2/sde1/rdr6* gene of *Arabidopsis thaliana* (Dalmay et al. 2000; Mourrain et al. 2000; Yu et al. 2003), the *ego-1* and *rrf-1* genes of *C. elegans* (Smardon et al. 2000; Sijen et al. 2001), and the *rrpA* gene of *Dictyostelium discoideum* (Martens et al. 2002) affect RNA silencing, demonstrating a clear genetic role for RdRPs in the mechanism of PTGS. *sgs2/sde1/rdr6* gene silencing mutants even proved to be a valuable genetic background for transformation with overexpression constructs, increasing the frequency of highly expressing transformants from 20 to 100% in wild-type and *sgs2* backgrounds, respectively (Butaye et al. 2004).

RdRPs may operate at multiple levels in RNA silencing. They might be implicated in the generation of sufficient initial trigger dsRNA from speculative aberrant transgenic RNAs. In line with this role, RdRPs are no longer required when a preformed dsRNA is expressed from inverted repeat (IR) constructs in plants and fungi (Béclin et al. 2002; Catalanotto et al. 2004), or when silencing is induced by viruses that encode their own RdRP proteins (Dalmay et al. 2000). On the other hand, in *C. elegans* (Sijen et al. 2001), *Dictyostelium* (Martens et al. 2002) and fission yeast (Schramke and Allshire 2003) the RdRP function is also required when the trigger dsRNA is delivered exogenously or expressed directly from hairpin constructs, suggesting that RdRPs might participate in more than just the production of the dsRNA trigger, and that they are involved in amplification of the silencing signal leading to very efficient RNA silencing throughout the organism. The extent to which RdRP action contributes to RNA silencing may depend on the organism or tissue, the specific induction pathway, or the target.

3
RNA Amplification and Transitive Silencing

Amplification of the RNA silencing signal by an RdRP could occur either by replicating the dsRNA trigger or by expanding the initial pool of siRNAs. The latter possibility is favored, because many studies support a model in which the primary siRNAs that are derived from the inducer dsRNA through Dicer activity serve as primers or are recognized as tags for synthesis of dsRNA by an RdRP using the target transcript RNA as a template. Subsequent cleavage of the dsRNA produced by Dicer results in the formation of new secondary siRNAs, in turn capable of targeting homologous target mRNAs for degradation. This RdRP/Dicer pathway can also lead to the production of siRNAs corresponding to sequences located outside the region of homology between the silencing inducer and the primary target, thus resulting in the silencing of secondary targets that do not show any sequence homology to the initial silencing trigger. This kind of RNAi induced by secondary siRNAs spreading along the target gene was designated transitive RNAi in *C. elegans* and also occurs in plants and fungi.

3.1
Transitive Silencing in *C. elegans*

Analysis of RNA populations during RNAi in *C. elegans* demonstrated, in addition to the expected trigger-derived siRNAs, the existence of small RNAs that correspond to regions upstream of the region targeted by the inducing dsRNA (Sijen et al. 2001). The abundance of these secondary siRNAs appeared to decrease as a function of the distance from the primary trigger, and to become negligible at distances greater than a few hundred base pairs. To test whether these secondary siRNAs are capable of targeting degradation of homologous mRNAs, a transitive RNAi assay was carried out, in which two populations of target RNA were present in the cell: a primary target with a segment of homology to the dsRNA trigger, and a secondary target with no homology to the trigger, but to the second segment of the primary target. In one first experiment, the primary target encoded a nuclear-targeted green fluorescent protein (GFP)-LacZ fusion protein, and the secondary target a mitochondrially targeted GFP. Injection of dsRNA segments from *lacZ* into animals carrying both transgenes resulted in the reduction of both nuclear GFP-LacZ and mitochondrial GFP. A trigger that was located just 3' to the *gfp-lacZ* junction was most potent in the transitive RNAi assay. When the same *lacZ* dsRNA trigger was used, but the order of segments in the primary target was reversed (LacZ-GFP), no transitive silencing could be obtained, consistent with the fact that no siRNAs downstream of the targeted region could be detected. In another experiment, the primary and secondary targets were a *unc-22-gfp* transgene, and the endogenous *unc-22* gene, respectively. Injection of *gfp* dsRNA into animals expressing the *unc-22-gfp* transgene produced the twitching phenotype that is characteristic of loss of *unc-22* expression. To test whether transitive RNAi could proceed with endogenous genes as primary and secondary targets, animals heterozygous for a functional deletion allele of *unc-22* were injected with dsRNA corresponding to the deleted region, which led to transitive silencing of both the wild-type and the deletion allele.

Similar results were obtained by Alder et al. (2003), who could induce silencing of three different endogenous genes in *C. elegans* by transitive RNAi assays. Directionality for the primary mRNA target was observed: transitive RNAi targets mRNA sequences located 5′ to the mRNA sequences homologous to the incoming dsRNA. The presence of the primary target mRNA is essential for the transitive effect, because it is presumed to function as a template for dsRNA synthesis. The most plausible explanation for the observed 3′–5′ directionality in *C. elegans* is that the siRNA functions as a primer to initiate elongation (Fig. 1; Lipardi et al. 2001). Alternatively, differential stability of potential templates derived from a cleaved target mRNA could account for directionality. The 5′ fragment may be stabler and therefore more frequently used as a template than the 3′ fragment (Han and Grierson 2002a). Finally, the machinery for RNA synthesis may be associated with the RISC and positioned for unidirectional RNA synthesis by this complex (Alder et al. 2003).

Interestingly, spreading could also be observed at the level of the dsRNA trigger (Alder et al. 2003). A dsRNA hairpin consisting of a *gfp* stem and a *unc-22* loop is capable of inducing a Unc-22 phenotype only when the hairpin trigger is continuously administered and a *gfp* mRNA target is present. This observation implies that *unc-22* dsRNA is synthesized de novo via transitive RNAi from the hairpin construct itself by priming of cleaved *gfp* mRNA or *gfp* siRNAs with the hairpin dsRNA as a template. In the case of siRNA priming, the *gfp* transgene might be necessary to produce sufficient secondary siRNAs or structurally modified siRNAs, which could act as primers.

C. elegans has four RdRP genes, *ego-1*, *rrf-1*, *rrf-2*, and *rrf-3*. Smardon et al. (2000) demonstrated that *ego-1* is required for efficient RNAi in the adult germline. When the role of the other three RdRPs in the transitive silencing process was investigated, *rrf-2* and *rrf-3* mutants were found sensitive to RNAi in all tissues (soma and germline) for both standard and transitive RNAi assays (Sijen et al. 2001). Interestingly, the *rrf-3* deletion strain showed increased sensitivity to RNAi when compared with that in wild-type animals, perhaps because of competition with other RdRPs (RRF-1 and EGO-1) for components or intermediates in the RNAi reaction (Simmer et al. 2002). By contrast, in *rrf-1* mutants, complete resistance against certain RNAi triggers in somatic tissue was observed, while interference was retained for genes expressed in the germline. Secondary siRNAs were not produced anymore, but the primary siRNAs corresponding to the original trigger RNA were still detected in *rrf-1* mutants. The lack of RNAi in the presence of primary siRNAs could have several reasons. First, the levels of primary siRNAs could be insufficient to trigger an efficient silencing response, perhaps because primary si-RNAs are less efficiently incorporated into the RISC than the secondary si-RNAs. Second, the initial si-RNA–mRNA interaction may be relatively transient or unstable in vivo and polymerization by an RdRP could be required for stabilization. Third, RRF-1 and other RdRPs could have an additional function in the RNAi pathway, for example, phosphorolyzing and breaking down the target RNA, or tagging it for destruction.

Additional support for an involvement of RdRPs in RNAi in *C. elegans* has been provided by Tijsterman et al. (2002), who showed that a broad range (22–40 nucleotides) of short antisense RNAs (asRNAs) can also efficiently trigger RNAi when injected in close proximity to the target mRNA. Because the asRNAs have less strict size requirements than double-stranded siRNAs, which are only functional when they are 21–23 nucleotides in length (Elbashir et al. 2001b; Nykänen et al. 2001), the asRNAs might enter the RdRP/Dicer pathway, instead of the RISC-mediated mRNA degradation pathway. Like siRNAs, the asRNAs could act as primers for an RdRP elongation reaction with the target RNA as a template, leading to dsRNA molecules that can subsequently be a substrate for Dicer action (Tijsterman et al. 2002). Consistently, modification of the 3′ ends of the asRNAs severely reduced the silencing efficiency of the asRNAs (Tijsterman et al. 2002), as observed in *Drosophila* embryo extracts, where double-stranded siRNAs were no longer incorporated into dsRNA when the 3′ hydroxyl group was blocked by a phosphate group (Lipardi et al. 2001). Because RDE-1 and RDE-4, which are required to initiate RNAi induced by injection of dsRNA or by transgene-produced dsRNA (Grishok et al. 2000), are dispensable for RNAi triggered by the asRNAs, the RdRP-generated dsRNAs might be delivered to Dicer in a manner different from that of exogenous or transgenic dsRNA (Tabara et al. 2002).

3.2
Transitive Silencing in Plants

Although the term transitive RNAi was first used for *C. elegans*, the phenomenon was initially observed in plants. Tomato transformants with extra copies of the tomato 1-aminocyclopropane-1-carboxylate (ACC) oxidase gene (*aco1*) carrying an IR of the 5′ untranslated region (UTR) showed co-suppression of both the transgenic and endogenous *aco1* genes and, in addition, also of the endogenous *aco2* gene. This latter gene only exhibits significant similarity to the *aco1* transgene in the coding region (Hamilton et al. 1998). This observation implies a spreading of the target region along the transgenic DNA or RNA from the IR in the 5′UTR to the downstream coding sequence. This explanation was supported by the detection of 5′UTR siRNAs and siRNAs corresponding to the region immediately downstream of the IR (Han and Grierson 2002b). In transgenic *gfp*-expressing *Nicotiana benthamiana* plants, bombardment of 5′ (GF) or 3′ (P) fragments of *gfp* complemenatary DNA induced not only systemic silencing of the integrated *gfp* transgene, but also led to targeting of nonoverlapping *gfp* sequences that were independently expressed from a virus-based vector (PVX-P and PVX-GF, respectively; Voinnet et al. 1998). These data demonstrate spreading of the PTGS target region in both 5′–3′ and 3′–5′ directions. Also virus-induced gene silencing (VIGS) of transgenes is associated with target-site spreading (Vaistij et al. 2002). *N. benthamiana* plants expressing a chimeric *gfp-nos* transgene were first infected with three different viral vectors carrying the 5′ or 3′ half of the *gfp* coding region, or the 3′*nos* sequence (TRV : GF, TRV : P, or TRV : NOS). These vectors induced efficient PTGS of the *gfp-nos* transgene and the production of GF-, P-, and NOS-siRNAs in all three plants. These plants were challenge-inoculated with an unrelated viral vector carrying the 3′ half of the *gfp* coding region (PVX : P). This unrelated vector accumulated to low levels in all three plants, indicating that the P region of *gfp* was a target irrespective of whether the initiator of RNA silencing was GF, P, or NOS. Spreading occurred from the initiator region in both 3′ (from GF to P) and 5′ (from NOS to P) directions, extending at least through 332 nucleotides (P region), and depending on RdRP (SDE1/SGS2/RDR6) function and on transcription of the *gfp-nos* transgene.

In contrast, VIGS of the endogenous genes phytoene desaturase (PDS) or ribulose-1,5-biphosphate carboxylase/oxygenase (Rubisco) did not lead to resistance against the challenge inoculum (Vaistij et al. 2002). miRNA-directed cleavage of endogenous mRNA targets is also not associated with transitive RNA silencing, even though miRNAs act like siRNAs in wheat extracts (Tang et al. 2003). Han and Grierson (2002a) showed that during co-suppression of an endogenous polygalacturonase gene and a 3′ truncated homologous transgene, the siRNAs were most probably produced from the transgene, because no siRNAs from the 3′ part of the endogenous transcripts were detected. These studies suggest that endogenous target RNAs are incapable of entering the amplification pathway, because of possible quantitative or qualitative restrictions. Perhaps the endogenous transcripts are not numerous enough to serve as templates (Tang et al. 2003), or SDE1/SGS2/RDR6 is unable to recognize them as templates because the endogenous RNAs lack certain characteristics or are associated with proteins inhibiting RdRP. In contrast, Sanders et al. (2002) have shown that endogenous glucanase genes play an active role in the synthesis of siRNAs in tobacco. During co-suppression of the endogenous *glb* gene and the *N. plumbaginifolia gn1* transgene (81% homology) small sense RNAs and asRNAs homologous to the endogenous *glb* gene were detected, indicating that the co-suppressed endogenous gene is involved in signal amplification and target selection.

The apparent discrepancy between the results of Sanders et al. (2002) and Vaistij et al. (2002) could be explained by the difference in silencing inducer:

in comparison with viral RNAs, transgenes could more efficiently recruit endogenous targets into an active role in silencing, perhaps involving a nuclear phase that is absent during VIGS. In the experiments of Han and Grierson (2002a), it could be that the endogenous transcripts do serve as templates, after cleavage by the RISC, and that the 5' half of the cleaved transcript, which is more abundant and perhaps stabler, is used preferentially as a template. Some endogenous genes might also be more susceptible to transitivity than others, because of differences in transcript stability, abundance, or secondary structure. However, this may not only be the case for endogenous genes, but also for transgenes, since transgenic RNA silencing systems that do not exhibit spreading of RNA targeting have been reported (Vogt et al. 2004; Wang et al. 2001). During silencing of a chimeric *gus-Sat* transcript induced by replicating satellite RNA only satellite-derived, but no *gus*-specific si-RNAs could be detected (Wang et al. 2001). In viroid-induced RNA silencing, potato spindle tuber viroid (PSTVd) infection led to silencing of a chimeric *gfp* transgene containing part of the PSTVd sequence, but not of a simultaneously expressed *gfp* transgene without any PSTVd homology (Vogt et al. 2004). This is striking since the *gfp* gene had been shown previously to be highly susceptible to transitive silencing in both the 5' and the 3' directions (Vaistij et al. 2002; Voinnet et al. 1998). Because only PSTVd-specific si-RNAs were detected, the inserted viroid sequence is believed to form certain secondary structures that block RdRP-mediated dsRNA synthesis from the chimeric *gfp-PSTVd* transcripts (Vogt et al. 2004).

Another report on spreading of the target region in *N. benthamiana* plants undergoing VIGS demonstrated a predominance of 5'–3' spreading over 3'–5' spreading, suggesting a link between the commonly observed preferential targeting of 3' regions of silenced transgenes and the RdRP-mediated spread of silencing (Braunstein et al. 2002; English et al. 1996; Hutvágner et al. 2000; Sijen et al. 1996). Small RNAs from the 3' region of *gus* could be observed after VIGS induction with PVX carrying a 5' *gus* sequence, but not vice versa (Braunstein et al. 2002). This predominance could result from disturbance or inhibition of RdRP function during unprimed elongation that started from the 3' end by certain features of the *gus* transcript in the 5' region, such as secondary structure. Alternatively, its cause might simply be the limited processivity of SGS2/SDE1/RDR6, allowing the polymerization of RNA only for a distance of a few hundred nucleotides from the 3' end. Accordingly, Van Houdt et al. (2003) detected in a transitive silencing system in tobacco only *gus*-specific small RNAs derived from the 3' part of a *gus* transcript. In this system, a post-transcriptionally silenced IR transgene locus X (*nptII-3'chs* IR) can trigger silencing of a nonhomologous transgene Z (*gus-3'nos*) when a stepwise homology is created through introduction of a chimeric primary target Y (*gus-3'chs*) with one region homologous to the silencing inducer X (3'*chs*) and a second upstream region homologous to the secondary target Z (*gus*). Small *gus*-specific RNAs accumulate only upon silencing of the *gus* genes in the presence of locus X. These results fit the hypothesis of RdRP-dependent synthesis of *gus* asRNA, primed by locus X-derived 3'*chs*-specific small RNAs on a locus-Y-derived transcript. Alternatively, mere association of the primary target transcript with an siRNA-protein complex or partial RISC-mediated degradation could allow the transcript to be recognized by an RdRP as a template for dsRNA synthesis (Fig. 1; Van Houdt et al. 2003). This primer-independent synthesis can account for the observed 5'–3' spreading in plants; the tomato RdRP is able to start polymerization preferentially at the 3' terminus of the template in the absence of any primer (Schiebel et al. 1993). This mechanism could also account for the observed spreading of targeting in tobacco BY2 protoplasts: co-transfection with a *gfp* plasmid and siRNAs targeted to the 5' region of the *gfp* gene sequence demonstrated de novo synthesis of siRNAs corresponding to the 3' region of the *gfp* target (Vanitharani et al. 2003).

3.3
Transitive Silencing in Fungi

Work on *N. crassa* provided the first evidence that purified recombinant QDE-1, a homologue of the tomato RdRP, possesses in vitro de novo and primer-dependent RNA polymerase activity in the absence of other cellular factors (Makeyev and Bamford 2002). The fact that QDE-1 is no longer required upon the direct expression of a hairpin dsRNA (Catalanotto et al. 2004) suggests that the main role of RdRP in gene silencing in *N. crassa* (quelling) is the production of sufficient dsRNA from the transgenic trigger RNA, which leads to the production and amplification of siRNAs. In line with this hypothesis, overexpression of QDE-1 results in an increase in the production of siRNAs and a concomitant dramatic increase in efficiency of quelling induced by transgene tandem repeats, indicating that QDE-1 is a rate-limiting factor in quelling (Forrest et al. 2004). In overexpressed strains, the number of transgene copies required to induce quelling is significantly reduced, suggesting that every transgenic/repetitive locus possesses the ability to produce a silencing signal, but that this signal is not always sufficient to induce silencing, because the trigger RNA is expressed below a certain threshold. Silencing activation and maintenance in *N. crassa* appear to rely on the amount of both QDE-1 and transgenic copies that produce trigger RNA molecules to be converted to dsRNA by the RdRP (Forrest et al. 2004). Nevertheless, these results do not exclude the possibility of target RNAs being used as templates for siRNA amplification, with transitive silencing in *N. crassa* as a result. On the other hand, an in vivo study in *Dictyostelium discoideum* showed that target-dependent amplification of small RNAs by the RdRP homologue RrpA is indispensable for hairpin-RNA-induced silencing, which requires detectable amounts of small RNAs (Martens et al. 2002). In the filamentous fungus *Mucor circinelloides*, silencing of the endogenous *carB* gene induced by an episomal 3′ truncated transgene was associated with the production of small sense RNAs and asRNAs, also corresponding to sequences downstream of the transgenic inducer sequence (Nicolás et al. 2003). The asRNAs were generated preferentially from the 3′ region of the endogenous *carB* transcript, whereas asRNAs from the 5′ region were hardly detectable, indicating that the majority of detected small RNAs are produced via primer-independent amplification by the putative RdRP from the 3′ end of the endogenous target mRNA, with low processivity.

4
RNA Amplification and Systemic Silencing

The observation that mRNAs are not only targets but also amplifiers of the initial silencing signal may not only explain the remarkable potency and observed transitivity of RNA silencing, but also the systemic propagation of RNA silencing throughout the organism. RNA silencing in plants and *C. elegans* is noncell autonomous: it can be induced locally and then spread from the initiation site to more distant sites, and even persist a long time after the original source of silencing has been eliminated. In plants, two types of transmission occur: short-range cell-to-cell movement through plasmodesmata, and bidirectional long-distance movement via phloem. Systemic silencing not only requires a system to pass the silencing signal from cell to cell, but also a mechanism to perceive and amplify the incoming signal.

4.1
Systemic Silencing in Plants

The first hint for the systemic nature of RNA silencing came from studies in which nonclonal spatial patterns of co-suppression were observed: spontaneous induction of PTGS started on a localized area of a single leaf and then propagated through the whole plant as a gradual process (Boerjan et al. 1994; Palauqui et al. 1996). Direct evidence for a systemic silencing signal was obtained by grafting experiments (Palauqui et al. 1997; Palauqui and Vaucheret 1998; Voinnet et al. 1998) and local delivery of exogenous DNA via *Agrobacterium tumefaciens* infiltration (Voinnet and Baulcombe 1997; Voinnet et al. 1998) or biolistics (Palauqui and Balzergue 1999; Voinnet and Baulcombe 1997). Grafting of the upper part (scion) of nonsilenced plants onto the lower part (stock) of silenced plants has demonstrated that the silencing signal can be transmitted across a graft junction and induce silencing in the scion. The signal produced in the silenced stock can even pass through part of a stem of a wild-type plant in which no homologous target RNA is present (Voinnet et al. 1998). Systemic co-suppression could only be observed when a transcriptionally active target transgene was present in the receiving cells, or when the endogenous target mRNA in the scion accumulated to high levels (Palauqui et al. 1997; Palauqui and Vaucheret 1998). These results suggest that the silencing signal can travel some distance through the phloem without any need for amplification, while in the receiving cells target-dependent amplification of the signal is required to obtain efficient silencing. Such amplification has been demonstrated during systemic PTGS induced by biolistically delivered siRNAs (Klahre et al. 2002). New siRNAs representing parts of the target RNA that are outside the region of homology with the triggering siRNA could be detected in systemically silenced new leaves, indicating that siRNAs themselves or intermediates induced by siRNAs could comprise systemic silencing signals and that these signals can initiate an amplification cycle in the receiving cells. As mentioned in Sect. 3.2, Voinnet et al. (1998) also observed spreading of the target region during systemic silencing induced by bombardment of fragments of GFP, which can be explained by target-dependent amplification of the silencing signal. Another study demonstrated that the progression of locally induced silencing to systemic silencing is determined by the ability of the receiving cells to propagate the signal: only plants capable of triggering spontaneous silencing of a transgene and the homologous endogenous gene showed biolistic activation of systemic co-suppression (Palauqui and Balzergue 1999). Also in grafting experiments only such plants were able to maintain the systemically induced co-suppression after grafting onto a wild-type stock (Palauqui and Vaucheret 1998).

Himber et al. (2003) studied the cell-to-cell movement of RNA silencing in more detail. *Agro*-infiltration of a *gfp* expressing the *N. benthamiana* line consistently produced a fine red border over a constant number of cells (10–15 cells) in which *gfp* silencing was triggered by a signal originating from the infiltrated area, independent of the presence of homologous transcripts. Simultaneously induced silencing of a *gfp* transgene and an endogenous Rubisco small subunit (*RbcS*) gene by a chimeric phloem-restricted viral vector (PVX-GP:RbcS:P-Δ25) revealed two types of cell-to-cell movement: limited movement of silencing of the endogenous gene *RbcS*, which was restricted to 10–15 cells around the veins, and extensive movement of *gfp* silencing, which progressively invaded the entire lamina. This observation could be explained by a differential capacity of the target mRNAs to sustain transitivity (Vaistij et al. 2002), since a similar experiment in *A. thaliana sde1* and *sde3* mutants showed that the extensive movement of *gfp* silencing depends on SDE1/SGS2/RDR6 and, to a lesser extent, on SDE3, an RNA helicase believed to be involved in SDE1-mediated dsRNA synthesis, whereas limited movement did not require either of those two proteins (Himber et al. 2003). This hypothesis was confirmed by molecular analysis of *sde1* and *sde3* mutants: the production of secondary siRNAs (21 nucleotides) through the action of SDE1 and SDE3 was necessary for extensive, but not for limited, movement, which relied only on the presence of primary siRNAs (21 and 25 nucleotides) that did not spread outside the region of homology with the hairpin trigger. Experiments with the viral suppressor P1 (Voinnet et al. 1999) showed that the 21-nucleotide siRNAs were sufficient for short-range movement, and limited movement of *RbcS* gene silencing was associated with production of mainly 21-nucleotide siRNAs (Himber et al. 2003). Therefore, these molecules are the most probable candidates for the short-distance signal molecule, which is consistent with the fact that synthetic siRNAs can induce systemic silencing (Klahre et al. 2002). On the basis of these results, a possible model for cell-to-cell movement of RNA silencing in plants has been proposed. Local initiation of silencing by *Agro*-infiltration, tissue-restricted VIGS, or tissue-specific expression of a hairpin RNA would produce 21- and 25-nucleotide primary siRNAs, but mainly the 21-nucleotide siRNAs would move to adjacent cells, without any need for target RNAs. The outcome of this initial wave of limited movement depends on the presence of homologous transcripts capable of serving as templates for the synthesis of secondary 21-nucleotide siRNAs through the action of SDE1 and SDE3. Like the primary siRNAs, these newly produced secondary siRNAs could create a new wave of limited movement to adjacent cells, where they initiate the same SDE1/SDE3-mediated process. Such reiteration of limited movement waves would eventually translate into extensive movement of RNA silencing (Himber et al. 2003). In accordance with this hypothesis, systemic disease resistance to plum pox virus, conferred by phloem-specific expression of hairpin dsRNA, is also associated with the RdRP-mediated production of secondary siRNAs using the inoculated viral genome as a template (Pandolfini al. 2003). Hamilton et al. (2002) showed that the 21-nucleotide siRNAs are far more abundant than the 25-nucleotide siRNAs in systemic tissues undergoing extensive cell-to-cell silencing movement; however, the onset of systemic silencing in newly emerging leaves was associated with the 25-nucleotide siRNAs. This suggests that there are separate signal molecules for cell-to-cell and long-distance transport. The 25-nucleotide siRNAs (or a derivative/precursor molecule) could act as a phloem-specific silencing signal, whereas the 21-nucleotide siRNAs are the short-distance signal molecules (Himber et al. 2003). However, other reports reveal no consistent correlation between the capacity for systemic silencing and the accumulation of any particular class of small RNA (García-Pérez et al. 2004; Mallory et al. 2003). Several studies support the existence of separate mechanisms for cell-to-cell and phloem transport of RNA silencing. First, some silencing suppressors have contrasting effects on each transport process (Himber et al. 2003). Second, systemic, but not local, spread of silencing can be blocked in the presence of low levels of cadmium (Ueki and Citovsky 2001). Third, a recent study showed that cell-to-cell RNA silencing movement can be independent of long-distance spread, which requires induction of RNA silencing in different cell types, including phloem cells (Ryabov et al. 2004).

Another report that combined transitive and systemic silencing processes supported the suggestion that some kind of RNA amplification product is the systemic silencing signal (García-Pérez et al. 2004). To investigate whether a primary target is capable of producing systemic silencing signals, a transitive silencing XYZ setup was used similar to that of Van Houdt et al. (2003), but instead of the three loci being present in the same plant genome, graftings were made: plants homozygous for the secondary target locus Z (*gus*-expressing ZZ scions) were grafted onto plants containing the silencer locus X and the primary target locus Y in different zygosity combinations (*gus*-silenced XXYY, XXY_, X_YY, and X_Y_ stocks). Induction of systemic *gus*-silencing was dosage-dependent: only a double dose of silencer X and/or of primary target Y in the stocks (XXYY, XXY_, and X_YY) resulted in systemic silencing in the ZZ scions. Secondary target Z in the scion did not show any transcript homology with silencing inducer X in the stock, indicating that a systemic silencing signal is not necessarily produced by the silencing trigger, but can also be generated from the primary target. Secondary siRNAs in XXY_ and X_Y_ stocks, which are capable and incapable of sending a systemic signal to the scion, respectively, have similar accumulation levels, suggesting that the mobile silencing signal does not comprise the secondary siRNAs themselves, but rather some other RNA amplification product. Such a product may be synthesized by an RdRP using as a template the transcripts derived from locus Y that are in some way marked, for instance, by the 3′*chs* siRNAs derived from locus X. This marking may be a limiting factor to produce sufficient signal for systemic silencing, which might explain the observed dosage dependence: doubling the amount of 3′*chs* siRNAs and/or of *gus* mRNAs via homozygosity of locus X and/or locus Y increases the probability of both molecules finding each other. To activate systemic silencing not only amplification in the rootstock is required to produce enough systemic silencing signal, but also amplification in the receiving scion to generate target-specific tertiary siRNAs that are able to establish silencing of the secondary target, as mentioned before (Sect. 3.2; Palauqui and Vaucheret 1998; Voinnet et al. 1998). For the latter process the presence of the secondary target gene is needed, indicating that the target transcripts are involved in the perception and/or subsequent conversion of the systemic transitive signal into a pool of tertiary siRNAs by target-dependent amplification (García-Pérez et al. 2004).

Consistent with the hypothesis of RNA amplification products being the mobile silencing signal is the fact that transgenic tobacco with different silencing-inducing loci vary in their capacity to trigger systemic RNA silencing (Mallory et al. 2003). A transgene IR-silenced line, which is able to induce systemic silencing, could efficiently enter the amplification pathway. It is hypothesized that there are sufficient target mRNAs that can serve as templates in response to occasionally produced dsRNA from the transgenes integrated as an IR. In contrast, two different amplicon-silenced lines fail to effectively transmit the systemic silencing signal; presumably no amplification can be initiated, because the target RNAs are efficiently degraded by rapidly assembled RISCs that are guided by siRNAs derived from the continuously produced dsRNA trigger (amplicon). A study in which RNA extracts prepared from *gfp*-silenced plants led to systemic RNAi after injection into *gfp*-expressing *C. elegans* revealed that an RNA of approximately 85 nucleotides was most active in inducing silencing in the worm (Boutla et al. 2002). This molecule, perhaps representing an RNA amplification product, could be responsible for the systemic spread in plants. Nonetheless, there is still much speculation over the nature of the systemic silencing signal(s) in plants.

4.2
Systemic Silencing in *C. elegans*

In *C. elegans*, RNAi is also not restricted to those cells that are exposed to a dsRNA trigger, and can even be transmitted to progeny (Fire et al. 1998). Systemic responses are observed by injecting dsRNA into any site of the ne- matodes (Fire et al. 1998), by soaking them in a solution containing dsRNAs (Maeda et al. 2001; Tabara et al. 1998), or by feeding them with dsRNA- expressing bacteria (Timmons and Fire 1998; Timmons et al. 2001). Also tissue-specific expression of a hairpin dsRNA trigger can lead to gene silenc- ing in other tissues (Winston et al. 2002), but this effect strongly depends upon the choice of promoter used to drive dsRNA expression. Worms that did not exhibit comprehensive systemic RNAi phenotypes in other tissues, did re- lease an RNA silencing signal from the cells expressing the dsRNA to more distant tissues after treatment with exogenous, unrelated dsRNA, demon- strating that dsRNA derived from the environment can not only trigger, but can also influence RNA-silencing mechanisms in nematodes (Timmons et al. 2003). In this manner, systemic RNAi may be part of a general mechanism for sensing and responding to environmental pathogens.

Genetic screens for *C. elegans* mutants defective in the systemic spread- ing of RNAi have been performed by two independent groups. Winston et al. (2002) isolated three 'systemic RNAi-deficient' (*sid*) mutants. The *sid-1* gene encodes a multispan transmembrane protein, enabling passive cellular uptake of dsRNA (Feinberg and Hunter 2003). *sid-1* mutants fail to exhibit systemic RNAi from transgene-derived and exogenously delivered dsRNAs, and show a reduced level of RNAi response in progeny of injected mutant animals (Win- ston et al. 2002). Tijsterman et al. (2004) identified five other 'RNAi spreading defective' (*rsd*) mutants from which the *rsd-8* gene is the same as the *sid-1* gene. These mutants could be divided into two classes: *rsd-4* and *rsd-8* mu- tants that are completely defective in the cellular uptake of dsRNA, and *rsd-2*, *rsd-3*, and *rsd-6* mutants that are not defective in the initial uptake of dsRNA from the gut into somatic tissues, but that are unable to further distribute this dsRNA to the germline.

As in plants, systemic RNAi in *C. elegans* also appears to involve transi- tive RNAi, but at the level of the trigger dsRNA. Introduction of a *gfp-unc-22* hairpin dsRNA in strains expressing *gfp* only in the digestive tract leads to efficient silencing of *unc-22*, which is exclusively expressed in the body wall muscles. This observation suggests that *unc-22* dsRNA, produced via transitive RNAi from the hairpin molecule, or some other RNA derivate, can move between cells (Alder et al. 2003). In contrast, injection experiments with small asRNAs failed to induce systemic RNAi, although they are believed to enter the amplification pathway (Tijsterman et al. 2002). More experiments are required to investigate the interplay between transitive and systemic RNAi in *C. elegans*.

5
Absence of Transitive and Systemic RNA Silencing in *Drosophila* and Mammals

RNAi in *Drosophila melanogaster* and mammals does not seem to involve the activity of an RdRP, since their genomes do not code for members of the RdRP family (Martinez et al. 2002; Schwarz et al. 2002). Furthermore, a variety of experiments argue against the need for RdRP-mediated amplification in the RNAi pathway of *Drosophila* and mammals. First, cordycepin, an inhibitor of RNA synthesis, does not prevent dsRNA-induced targeting of endogenous mRNAs in mouse oocytes and early embryos (Stein et al. 2003). Second, siRNA-mediated RNAi in human cells is transitory, with cells recovering from a single treatment in 4–6 days (Holen et al. 2002; Kisielow et al. 2002), suggesting that the original siRNAs are not amplified to sustain an RNAi response. Third, blocking the 3′ hydroxyl termini of siRNAs does not influence their ability to efficiently trigger RNA silencing in *Drosophila* embryo lysate and in human cells (Chiu and Rana 2002; Martinez et al. 2002; Schwarz et al. 2002). This observation is in contrast with the random degradative PCR model proposed upon the observation that siRNAs can act in vitro as primers for secondary dsRNA synthesis, for which the presence of both the 5′ phosphate group and the 3′ hydroxyl group is critical (Lipardi et al. 2001). However, the RdRP activity identified in *Drosophila* embryo extract could be unrelated to the RNAi mechanism. Fourth, dsRNA corresponding to an alternatively spliced exon selectively degrades specific alternatively spliced mRNA isoforms in cultured *Drosophila* cells, implicating the absence of transitive RNAi in *Drosophila* (Celotto and Graveley 2002). Fifth, transitive effects directed to sequences downstream or upstream of the initial trigger region were not observed (Roignant et al. 2003). Sixth, a re-creation of the transitive RNAi assay of Sijen et al. (2001) in human cells clearly showed a lack of transitive RNAi in these cells (Chi et al. 2003). In the mosquito *Anopheles gambiae*, RNAi also does not show spreading outside the region of the target region (Hoa et al. 2003). All these data imply that RdRP-mediated transitive RNAi may represent an ancient phenomenon that was lost in higher animals.

In contrast to plants and nematodes, RNAi in *Drosophila* shows a strict cell autonomy. Cell-specific expression of a dsRNA trigger failed to induce spreading of RNAi throughout the fruit fly, perhaps because certain components are missing, such as transporters to export the signal from cells undergoing the RNAi process and receptors for importing the signal in adjacent cells (Roignant et al. 2003). In support of this hypothesis is the fact that the *Drosophila* genome does not seem to code for homologues of the *sid-1* gene of *C. elegans* encoding a transmembrane protein involved in dsRNA uptake in cells (Winston et al. 2002). As demonstrated in plants, systemic RNAi might also require an amplification step to translate the mobile silencing signal into silencing–directing RNA molecules (si-RNAs) in distant cells. The absence of detectable systemic RNAi in *Drosophila* might simply reflect the lack of detectable transitive RNAi, which involves the de novo synthesis of secondary siRNAs. *sid-1* and *rsd-3* homologues have been found in humans and mice (Winston et al. 2002; Kennedy et al. 2004), suggesting that RNAi in mammals could be systemic, but this still has to be demonstrated.

Why RdRP-mediated amplification should be required for RNAi in some organisms, but not in others, is a question that remains difficult to answer. Transitive RNAi may have been lost in mammals because of the evolution of the interferon (IFN) response, which is induced by long dsRNA. This IFN response results in the general inhibition of translation by the protein kinase R pathway, aspecific degradation of mRNAs by an induced RNase L, and finally apoptosis (Stark et al. 1998), but the use of si-RNAs can overcome this response. The presumed absence of an amplification mechanism in mammals may ensure that no aspecific responses are induced. A second possibility is the differential stability of primary si-RNAs. In plants, nematodes, and fungi, si-RNAs might be unstable and RdRP-mediated formation of secondary si-RNAs could increase the amount of siRNA-loaded RISCs either by amplification of the original pool of si-RNAs or by different features of the secondary si-RNAs produced, leading to an increased stability or to more efficient incorporation into RISCs. Supporting this hypothesis is the fact that *rrf1* mutants show no detectable steady-state levels of si-RNAs in vivo, whereas they retain the ability to produce a small population of siRNAs in vitro by RdRP-independent cleavage of the original trigger (Sijen et al. 2001). A recent paper describes the existence of an 'siRNase' (ERI-1), identified in a genetic screen for *C. elegans* mutants with enhanced sensitivity to dsRNA (Kennedy et al. 2004). This ERI-1 protein and its human orthologue degrade si-RNAs in vitro; presumably by affecting their $3'$ overhangs, si-RNAs become either nonfunctional and unable to enter the RISC or unstable and subsequently degraded by additional nucleases. Negative regulation of RNAi by ERI-1 may limit the duration, cell type specificity, or endogenous function of RNAi, and might operate more efficiently in *C. elegans* than in human cells. In *Drosophila* and mammals, siRNAs might be incorporated into the RISC very rapidly and efficiently. In vitro experiments in human cell extracts show that the RISC is formed during the first 15 min of incubation and that the siRNAs are irreversibly associated with the protein components of the RISC (Martinez et al. 2002).

The fact that RNAi in *Drosophila* and mammals is not transitive or systemic makes it a powerful tool for reverse genetics at the resolution of a single mRNA isoform and of a single cell type. RNAi triggered against one specific gene in a gene family or against one specific splicing variant of a gene will be limited to that one target. When induced in one specific tissue, RNAi will not spread to other tissues.

Acknowledgements This work was supported by grants from the European Union Biotech Project (no. QLRT-2000-00078). A.B. is a Research Fellow of the Fund for Scientific Research (Flanders) and H.V.H. is indebted to the Instituut voor de aanmoediging van innovatie door Wetenschap en Technologie in Vlaanderen for a postdoctoral fellowship.

References

Alder MN, Sames S, Gaudet J, Mango SE (2003) Gene silencing in *Caenorhabditis elegans* by transitive RNA interference. RNA 9:25–32

Béclin C, Boutet S, Waterhouse P, Vaucheret H (2002) A branched pathway for transgene-induced RNA silencing in plants. Curr Biol 12:684–688

Bernstein E, Caudy AA, Hammond SM, Hannon GJ (2001) Role for a bidentate ribonuclease in the initiation step of RNA interference. Nature 409:363–366

Boerjan W, Bauw G, Van Montagu M, Inzé D (1994) Distinct phenotypes generated by overexpression and suppression of S-adenosyl-L-methionine synthetase reveal developmental patterns of gene silencing in tobacco. Plant Cell 6:1401–1414

Boutla A, Kalantidis K, Tavernarakis N, Tsagris M, Tabler M (2002) Induction of RNA interference in *Caenorhabditis elegans* by RNAs derived from plants exhibiting posttranscriptional gene silencing. Nucleic Acids Res 30:1688–1694

Braunstein TH, Moury B, Johannessen M, Albrechtsen M (2002) Specific degradation of 3′ regions of GUS mRNA in posttranscriptionally silenced tobacco lines may be related to 5′–3′ spreading of silencing. RNA 8:1034–1044

Butaye KMJ, Goderis IJWM, Wouters PFJ, Pues JM-TG, Delauré SL, Broekaert WF, Depicker A, Cammue BPA, De Bolle MFC (2004) Stable high-level transgene expression in *Arabidopsis thaliana* using gene silencing mutants and matrix attachment regions. Plant J 39:440–449

Catalanotto C, Pallotta M, ReFalo P, Sachs MS, Vayssie L, Macino G, Cogoni C (2004) Redundancy of the two Dicer genes in transgene-induced posttranscriptional gene silencing in *Neurospora crassa*. Mol Cell Biol 24:2536–2545

Celotto AM, Graveley BR (2002) Exon-specific RNAi: a tool for dissecting the functional relevance of alternative splicing. RNA 8:718–724

Chi J-T, Chang HY, Wang NN, Chang DS, Dunphy N, Brown PO (2003) Genomewide view of gene silencing by small interfering RNAs. Proc Natl Acad Sci USA 100:6343–6346

Chiu Y-L, Rana TM (2002) RNAi in human cells: basic structural and functional features of small interfering RNA. Mol Cell 10:549–561

Cogoni C, Macino G (1999) Gene silencing in *Neurospora crassa* requires a protein homologous to RNA-dependent RNA polymerase. Nature 399:166–169

Dalmay T, Hamilton A, Rudd S, Angell S, Baulcombe DC (2000) An RNA-dependent RNA polymerase gene in *Arabidopsis* is required for posttranscriptional gene silencing mediated by a transgene but not by a virus. Cell 101:543–553

Elbashir SM, Lendeckel W, Tuschl T (2001a) RNA interference is mediated by 21- and 22-nucleotide RNAs. Genes Dev 15:188–200

Elbashir SM, Martinez J, Patkaniowska A, Lendeckel W, Tuschl T (2001b) Functional anatomy of siRNAs for mediating efficient RNAi in *Drosophila melanogaster* embryo lysate. EMBO J 20:6877–6888

Feinberg EH, Hunter CP (2003) Transport of dsRNA into cells by the transmembrane protein SID-1. Science 301:1545–1547

Fire A, Xu S, Montgomery MK, Kostas SA, Driver SE, Mello CC (1998) Potent and specific genetic interference by double-stranded RNA in *Caenorhabditis elegans*. Nature 391:806–811

Forrest EC, Cogoni C, Macino G (2004) The RNA-dependent RNA polymerase, QDE-1, is a rate-limiting factor in post-transcriptional gene silencing in *Neurospora crassa*. Nucleic Acids Res 32:2123–2128

García-Pérez RD, Van Houdt H, Depicker A (2004) Spreading of post-transcriptional gene silencing along the target gene promotes systemic silencing. Plant J 38:594–602

Grishok A, Tabara H, Mello CC (2000) Genetic requirements for inheritance of RNAi in *C. elegans*. Science 287:2494–2497

Hamilton A, Voinnet O, Chappell L, Baulcombe D (2002) Two classes of short interfering RNA in RNA silencing. EMBO J 21:4671–4679

Hamilton AJ, Brown S, Yuanhai H, Ishizuka M, Lowe A, Alpuche Solis A-G, Grierson D (1998) A transgene with repeated DNA causes high frequency, post-transcriptional suppression of ACC-oxidase gene expression in tomato. Plant J 15:737–746

Hammond SM, Bernstein E, Beach D, Hannon GJ (2000) An RNA-directed nuclease mediates post-transcriptional gene silencing in *Drosophila* cells. Nature 404:293–296

Han Y, Grierson D (2002a) Relationship between small antisense RNAs and aberrant RNAs associated with sense transgene mediated gene silencing in tomato. Plant J 29:509–519

Han Y, Grierson D (2002b) The influence of inverted repeats on the production of small antisense RNAs involved in gene silencing. Mol Genet Genomics 267:629–635

Himber C, Dunoyer P, Moissiard G, Ritzenthaler C, Voinnet O (2003) Transitivity-dependent and -independent cell-to-cell movement of RNA silencing. EMBO J 22:4523–4533

Hoa NT, Keene KM, Olson KE, Zheng L (2003) Characterization of RNA interference in an *Anopheles gambiae* cell line. Insect Biochem Mol Biol 33:949–957

Holen T, Amarzguioui M, Wiiger MT, Babaie E, Prydz H (2002) Positional effects of short interfering RNAs targeting the human coagulation trigger Tissue Factor. Nucleic Acids Res 30:1757–1766

Hutvágner G, Mlynárová L, Nap J-P (2000) Detailed characterization of the posttranscriptional gene-silencing-related small RNA in a GUS gene-silenced tobacco. RNA 6:1445–1454

Kennedy S, Wang D, Ruvkun G (2004) A conserved siRNA-degrading RNase negatively regulates RNA interference in *C. elegans*. Nature 427:645–649

Kisielow M, Kleiner S, Nagasawa M, Faisal A, Nagamine Y (2002) Isoform-specific knockdown and expression of adapter protein ShcA using interfering RNA. Biochem J 363:1–5

Klahre U, Crété P, Leuenberger SA, Iglesias VA, Meins F Jr (2002) High molecular weight RNAs and small interfering RNAs induce systemic posttranscriptional gene silencing in plants. Proc Natl Acad Sci USA 99:11981–11986

Lipardi C, Wei Q, Paterson BM (2001) RNAi as random degradative PCR: siRNA primers convert mRNA into dsRNAs that are degraded to generate new siRNAs. Cell 107:297–307

Maeda I, Kohara Y, Yamamoto M, Sugimoto A (2001) Large-scale analysis of gene function in *Caenorhabditis elegans* by high-throughput RNAi. Curr Biol 11:171–176

Makeyev EV, Bamford DH (2002) Cellular RNA-dependent RNA polymerase involved in posttranscriptional gene silencing has two distinct activity modes. Mol Cell 10:1417–1427

Mallory AC, Mlotshwa S, Bowman LH, Vance VB (2003) The capacity of transgenic tobacco to send a systemic RNA silencing signal depends on the nature of the inducing transgene locus. Plant J 35:82–92

Martens H, Novotny J, Oberstrass J, Steck TL, Postlethwait P, Nellen W (2002) RNAi in *Dictyostelium*: the role of RNA-directed RNA polymerases and double-stranded RNase. Mol Biol Cell 13:445–453

Martinez J, Patkaniowska A, Urlaub H, Lührmann R, Tuschl T (2002) Single-stranded antisense siRNAs guide target RNA cleavage in RNAi. Cell 110:563–574

Mourrain P, Béclin C, Elmayan T, Feuerbach F, Godon C, Morel J-B, Jouette D, Lacombe A-M, Nikic S, Picault N, Rémoué K, Sanial M, Vo T-A, Vaucheret H (2000) *Arabidopsis SGS3* and *SGS3* genes are required for posttranscriptional gene silencing and natural virus resistance. Cell 101:533–542

Nicolás FE, Torres-Martínez S, Ruiz-Vázquez RM (2003) Two classes of small antisense RNAs in fungal RNA silencing triggered by non-integrative transgenes. EMBO J 22:3983–3991

Nykänen A, Haley B, Zamore PD (2001) ATP requirements and small interfering RNA structure in the RNA interference pathway. Cell 107:309–321

Palauqui J-C, Balzergue S (1999) Activation of systemic acquired silencing by localised introduction of DNA. Curr Biol 9:59–66

Palauqui J-C, Vaucheret H (1998) Transgenes are dispensable for the RNA degradation step of cosuppression. Proc Natl Acad Sci USA 95:9675–9680

Palauqui J-C, Elmayan T, Dorlhac de Borne F, Crété P, Charles C, Vaucheret H (1996) Frequencies, timing, and spatial patterns of co-suppression of nitrate reductase and nitrite reductase in transgenic tobacco plants. Plant Physiol 112:1447–1456

Palauqui J-C, Elmayan T, Pollien J-M, Vaucheret H (1997) Systemic acquired silencing: transgene-specific post-transcriptional silencing is transmitted by grafting from silenced stocks to non-silenced scions. EMBO J 16:4738–4745

Pandolfini T, Molesini B, Avesani L, Spena A, Polverati A (2003) Expression of self-complementary hairpin RNA under the control of the *rolC* promoter confers systemic disease resistance to plum pox virus without preventing local infection. BMC Biotechnol 3:71–715

Roignant J-Y, Carré C, Mugat B, Szymczak D, Lepesant J-A, Antoniewski C (2003) Absence of transitive and systemic pathways allows cell-specific and isoform-specific RNAi in *Drosophila*. RNA 9:299–308

Ryabov EV, van Wezel R, Walsh J, Hong Y (2004) Cell-to-cell, but not long-distance, spread of RNA silencing that is induced in individual epidermal cells. J Virol 78:3149–3154

Sanders M, Maddelein W, Depicker A, Van Montagu M, Cornelissen M, Jacobs J (2002) An active role for endogenous β-1,3-glucanase genes in transgene-mediated cosuppression in tobacco. EMBO J 21:5824–2832

Schiebel W, Haas B, Marinkovic S, Klanner A, Sänger HL (1993a) RNA-directed RNA polymerase from tomato leaves. I. Purification and physical properties. J Biol Chem 263:11851–11857

Schiebel W, Haas B, Marinkovic S, Klanner A, Sänger HL (1993b) RNA-directed RNA polymerase from tomato leaves. II. Catalytic *in vitro* properties. J Biol Chem 268:11858–11867

Schiebel W, Pélissier T, Riedel L, Thalmeir S, Schiebel R, Kempe D, Lottspeich F, Sänger HL, Wassenegger M (1998) Isolation of an RNA-directed RNA polymerase-specific cDNA clone from tomato. Plant Cell 10:2087–2101

Schramke V, Allshire R (2003) Hairpin RNAs and retrotransposon LTRs effect RNAi and chromatin-based gene silencing. Science 301:1069–1074

Schwarz DS, Hutvágner G, Haley B, Zamore PD (2002) Evidence that siRNAs function as guides, not primers, in the *Drosophila* and human RNAi pathways. Mol Cell 10:537–548

Sijen T, Wellink J, Hiriart J-B, van Kammen A (1996) RNA-mediated virus resistance: role of repeated transgenes and delineation of targeted regions. Plant Cell 8:2277–2294

Sijen T, Fleenor J, Simmer F, Thijssen KL, Parrish S, Timmons L, Plasterk RHA, Fire A (2001) On the role of RNA amplification in dsRNA-triggered gene silencing. Cell 107:465–476

Simmer F, Tijsterman M, Parrish S, Koushika SP, Nonet ML, Fire A, Ahringer J, Plasterk RHA (2002) Loss of putative RNA-directed RNA polymerase RRF-3 makes *C. elegans* hypersensitive to RNAi. Curr Biol 12:1317–1319

Smardon A, Spoerke JM, Stacey SC, Klein ME, Mackin N, Maine EM (2000) EGO-1 is related to RNA-directed RNA polymerase and functions in germ-line development and RNA interference in *C. elegans*. Curr Biol 10:169–178; Erratum (2000) Curr Biol 10:R393

Stark GR, Kerr IM, Williams BRG, Silverman RH, Schreiber RD (1998) How cells respond to interferons. Annu Rev Biochem 67:227–264

Stein P, Svoboda P, Anger M, Schultz RM (2003) RNAi: mammalian oocytes do it without RNA-dependent RNA polymerase. RNA 9:187–192

Tabara H, Grishok A, Mello CC (1998) RNAi in *C. elegans*: soaking in the genome sequence. Science 282:430–431

Tabara H, Yigit E, Siomi H, Mello CC (2002) The dsRNA binding protein RDE-4 interacts with RDE-1, DCR-1, and a DExH-box helicase to direct RNAi in *C. elegans*. Cell 109:861–871

Tang G, Reinhart BJ, Bartel DP, Zamore PD (2003) A biochemical framework for RNA silencing in plants. Genes Dev 17:49–63

Tijsterman M, Ketting RF, Okihara KL, Sijen T, Plasterk RHA (2002) RNA helicase MUT-14-dependent gene silencing triggered in *C. elegans* by short antisense RNAs. Science 295:694–697

Tijsterman M, May RC, Simmer F, Okihara KL, Plasterk RHA (2004) Genes required for systemic RNA interference in *Caenorhabditis elegans*. Curr Biol 14:111–116

Timmons L, Fire A (1998) Specific interference by ingested dsRNA. Nature 395:854–854

Timmons L, Court DL, Fire A (2001) Ingestion of bacterially expressed dsRNA can produce specific and potent genetic interference in *Caenorhabditis elegans*. Gene 263:103–112

Timmons L, Tabara H, Mello CC, Fire AZ (2003) Inducible systemic RNA silencing in *Caenorhabditis elegans*. Mol Biol Cell 14:2972–2983

Ueki S, Citovsky V (2001) Inhibition of systemic onset of post-transcriptional gene silencing by non-toxic concentrations of cadmium. Plant J 28:283–291

Vaistij FE, Jones L, Baulcombe DC (2002) Spreading of RNA targeting and DNA methylation in RNA silencing requires transcription of the target gene and a putative RNA-dependent RNA polymerase. Plant Cell 14:857–867

Van Houdt H, Bleys A, Depicker A (2003) RNA target sequences promote spreading of RNA silencing. Plant Physiol 131:245–253

Vanitharani R, Chellappan P, Fauquet CM (2003) Short interfering RNA-mediated interference of gene expression and viral DNA accumulation in cultured plant cells. Proc Natl Acad Sci USA 100:9632–9636

Vogt U, Pélissier T, Pütz A, Razvi F, Fischer R, Wassenegger M (2004) Viroid-induced RNA silencing of GFP-viroid fusion transgenes does not induce extensive spreading of methylation or transitive silencing. Plant J 38:107–118

Voinnet O, Baulcombe DC (1997) Systemic signalling in gene silencing. Nature 389:553

Voinnet O, Vain P, Angell S, Baulcombe D (1998) Systemic spread of sequence-specific transgene RNA degradation in plants is initiated by localized introduction of ectopic promoterless DNA. Cell 95:177–187

Voinnet O, Pinto YM, Baulcombe DC (1999) Suppression of gene silencing: a general strategy used by diverse DNA and RNA viruses of plants. Proc Natl Acad Sci USA 96:14147–14152

Wang M-B, Wesley SV, Finnegan EJ, Smith NA, Waterhouse PM (2001) Replicating satellite RNA induces sequence-specific DNA methylation and truncated transcripts in plants. RNA 7:16–28

Winston WM, Molodowitch C, Hunter CP (2002) Systemic RNAi in C. elegans requires the putative transmembrane protein SID-1. Science 295:2456–2459

Yu D, Fan B, MacFarlane SA, Chen Z (2003) Analysis of the involvement of an inducible Arabidopsis RNA-dependent RNA polymerase in antiviral defense. Mol Plant-Microbe Interact 16:206–216

Zamore PD, Tuschl T, Sharp PA, Bartel DP (2000) RNAi: double-stranded RNA directs the ATP-dependent cleavage of mRNA at 21 to 23 nucleotide intervals. Cell 101:25–33

Nucleic Acids and Molecular Biology, Vol. 17
Wolfgang Nellen, Christian Hammann (Eds.)
Small RNAs
© Springer-Verlag Berlin Heidelberg 2005

RNA Interference and Antisense Mediated Gene Silencing

Markus Kuhlmann · Blaga Popova · Wolfgang Nellen (✉)

Abt. Genetik, Universität Kassel, Heinrich-Plett-Str. 40, 34132 Kassel, Germany
nellen@uni-kassel.de

Abstract Gene silencing by RNA interference (RNAi) and by antisense RNA are powerful tools to interfere with the expression of eukryotic genes. Since the first description
of RNAi in 1998, antisense-mediated gene silencing has been considered to have essentially the same mechanism as gene silencing by RNAi. However, while substantial effort
has been made to dissect the RNAi pathway, the cellular machinery that is responsible
for posttranscriptional regulation by antisense RNA is rather poorly defined and direct
comparisons between the RNAi and antisense experiments are rare. Even though similarities are very likely, recent data suggest that in addition to the expected overlaps in the
pathways, there are also mechanistic differences and different requirements for specific
gene products. We will summarize the current state of knowledge of the antisense RNA
and RNAi mechanisms and address some of the open questions in the field. We will further provide some evidence suggesting that gene silencing by antisense RNA and by RNAi
represent related but not identical mechanisms. A model to explain the partially overlapping pathways will be presented and may contribute to the further understanding of
posttranscriptional gene regulation.

1
Introduction

With the discovery of the first endogenous antisense RNAs (Mizuno et al.
1984) and the generation of antisense transgenes (Coleman et al. 1985), a new
era of interfering with gene expression was initiated (Brantl 2002). Unfortu-
nately, the expectations of this promising technology were only met in a very
limited number of cases. Problems occurred on different levels: (1) some
genes could not be silenced at all, (2) the degree of silencing varied considerably between cells and tissues within an experiment, and (3) silenced cells
tended to revert and lose silencing after prolonged growth. The rate of success to sufficiently knock down the expression of a specific gene was therefore
rather unpredictable. As a consequence, applications were mostly restricted
to basic science, where efficiently silenced cells could be selected for further
experiments and where even a low success rate was tolerable. The question
why cells even in a mostly homogeneous population may respond differently
to a challenge by antisense RNA is still not resolved.

The failure of antisense technology to achieve a mayor breakthrough was
probably due to the fact that researchers concentrated on, for example, med-

ical applications before understanding, in sufficient detail, the cellular mechanisms that lead to antisense-mediated gene silencing.

In 1998, the paper by Fire et al. (1998) revolutionized the field in several aspects: it specified the biologically active component as double-stranded RNA (dsRNA) and not an antisense transcript and, even more importantly, it initiated an unprecedented worldwide investigation of the cellular mechanisms involved in posttranscriptional gene silencing. This was exactly the kind of research that was lacking in the antisense field. At the same time, however, this first paper on RNA interference (RNAi) somehow implied that antisense-mediated gene silencing was either an indirect trigger for RNAi or that successful antisense experiments were due to minor amounts of "contaminating" dsRNA. Consequently, antisense research was not as vigorously pursued as before.

On the other hand, the discovery of RNAi-related endogenous mechanisms led to an increasing number of bioinformatics and experimental approaches in search for putative *cis-* or *trans*-encoded cellular RNAs that could form double strands and thus provide potential targets for the RNAi machinery. In addition, RNAi established a connection to the previously found heterochronic small antisense RNAs (Lee et al. 1993; Wightman et al. 1993) and laid ground for an unexpected new level of gene regulation by microRNAs. Furthermore, the RNAi pathway proved to be involved in epigenetic chromatin remodelling.

2
Parts of the RNAi Machine

Protein complexes involved in the RNAi mechanism have been rapidly unraveled within the last 5 years. By genetic as well as by biochemical assays, Dicer was defined as the nuclease that processes dsRNA to small interfering RNAs (siRNAs) of 21 base pairs with two nucleotide 3′ overhangs. Interestingly, a double-stranded RNase (dsRNase) activity that generated short processing products in *Dictyostelium* was implied in antisense-mediated gene silencing in 1989 (Möhrle 1989) and the participation of dsRNases of the RNase III family had previously been discussed in antisense-mediated gene silencing (Nellen and Lichtenstein 1993).

Bacterial RNase III contains a single dsRNase domain and acts as a homodimer that generates 9-base-pair fragments with two nucleotide 3′ overhangs from dsRNA (Blaszczyk et al. 2001). In contrast, Dicer has two C-terminal dsRNase domains that form an intramolecular pseudodimer. Each domain cuts one strand of the double helix. The central PAZ domain (named after the proteins Piwi, Argonaute and Zwille) and the C-terminal double-stranded RNA binding domain (dsRBD) apparently serve as a ruler to guide the cleavage reaction 21 base pairs away from the end of the double strand

(Zhang et al. 2004) and thus generate the 21-base-pair siRNAs that are indicative for RNAi. This model suggests that the enzyme initiates cleavage at the ends of a dsRNA and that it may be processive. In vitro, the activity of recombinant human Dicer is very low and processivity has not yet been shown. Probably this is due to the lack of other cellular factors that are required for turnover. Internal cleavage has been observed in vitro but it is not clear if this was due to a genuine internal dicing activity or to artificial entry sites generated by RNA damage.

At least in mammalian cells, dsRNA has for long been known to elicit a different response, namely, the activation of the dsRNA-dependent protein kinase R PKR (Pestka et al. 1987). PKR phosphorylates, among other targets, the translation initiation factor eIF2α and thereby leads to an overall inhibition of protein synthesis and eventually apoptosis. PKR and Dicer would compete for interaction with long dsRNA and obviously the PKR pathway would dominate because a single activated kinase molecule could initiate the signaling cascade to apoptosis; therefore, RNAi and the PKR response appear mutually exclusive. In fact, long dsRNAs introduced into mammalian cells result in the induction of interferon and related PKR-induced genes rather than in the specific elimination of a target messenger RNA mRNA (Billy et al. 2001). Since the tandemly repeated dsRBDs in PKR require at least 33 base pairs of dsRNA for activation (Schmedt et al. 1995), siRNAs are not sufficiently long to elicit the PKR response. Preformed siRNAs can therefore induce specific RNAi in mammalian cells (Elbashir et al. 2001); however, even short siRNAs have now been shown to elicit the PKR response under certain conditions (Persengiev et al. 2004; Sledz et al. 2003).

dsRNA may also be targeted by another cellular machinery: many organisms, including mammals and flies, contain the "adenosine deaminase that acts on RNA" (ADAR), an enzyme that converts adenosines to inosines in dsRNA. ADAR has specific targets in a few genes and contributes to protein diversity by modification of single codons (Bass 2002). In addition, however, ADAR also targets and edits long viral dsRNA, thus destroying its coding and replication capacity. Extensively edited dsRNA is not a substrate for Dicer and cannot be processed to 21-base-pair siRNAs. In *Caenorhabditis elegans*, ADARs apparently protect some intramolecular dsRNA from being targeted by Dicer in that A-to-I editing impairs its double-stranded characteristics (Tonkin and Bass 2003). This was concluded from an experiment where the phenotype of an ADAR mutant in *C. elegans* could be rescued by the secondary introduction of mutants in the RNAi pathway. However, the question arises how dsRNA targets discriminate between being diced or being edited and why transgene-derived dsRNA is not sufficiently edited to inhibit RNAi.

At least in some organisms, two or even three mechanisms may act on invading or endogenously generated dsRNA. So far, it appears that these different defense strategies are not closely co-ordinated but rather evolved independently.

In the RNAi pathway, the RNA-induced silencing complex (RISC) was identified as the second component that specifically picked up one strand of the siRNA from Dicer, transfered it to the target gene and mediated cleavage of the target opposite the center of the siRNA (Bernstein et al. 2001; Hammond et al. 2000, 2001). During RISC assembly, the siRNA is unwound and only one strand remains in the complex, while the other is degraded. The choice of strands is determined by the "easy 5'-end rule," i.e., the strand with the less stable base pairing at the 5' end is incorporated into the complex (Khvorova et al. 2003; Schwarz et al. 2003). R2D2, a dsRNA binding protein (Liu et al. 2003), provides a bridge between the initiator Dicer and the effector, the RISC. At least in *Drosophila*, the RISC is assembled via an intermediate complex that contains Dicer and an Argonaute protein (Lee et al. 2004; Pham et al. 2004). In the Dicer–R2D2 heterodimer, R2D2 binds to the 5'-phosphate of the more stable siRNA end. The orientation of the complex with respect to the siRNA thus determines which strand is transferred to the Argonaute protein AGO2, which is one of the key components in the RISC (Tomari et al. 2004).

The cleaving activity in the RISC had long been elusive until AGO2 (Liu et al. 2004; Meister et al. 2004) was found to have structural similarity to bacterial RNaseH (Song et al. 2004) and could be shown to serve as the "Slicer" nuclease that cuts the target opposite the center of the siRNA (Meister et al. 2004). The unprotected ends of the sliced mRNA are then rapidly degraded by the exosome and by Xrn related pathways (Orban and Izaurralde 2005). An unsolved problem is why only AGO2 appears to have slicing activity even though other Argonaute proteins have highly similar structures and the essential amino acids required for cleavage appear to be in the correct positions.

2.1
RNA-Directed RNA Polymerases

In most organisms, RNA-directed RNA polymerases (RdRPs) are required for RNAi. Usually they constitute a family of several related genes. A knockout or mutation of at least one specific member of the gene family results in a complete inability to respond to dsRNA by gene silencing. Cellular RdRPs were first detected in tomato (Schiebel et al. 1993a,b). Later, it was shown that at least some of these enzymes play an essential role in RNAi in *C. elegans* (Smardon et al. 2000), *Neurospora* (Cogoni and Macino 1999), *Arabidopsis* (Dalmay et al. 2000), *Dictyostelium* (Martens et al. 2002), *Schizosaccharomyces pombe* (Hall et al. 2002) and others.

In *C. elegans* as well as in plants, RdRPs are involved in transitive gene silencing. This describes the phenomenon that secondary siRNAs are produced from the 5' end of a chimeric construct when only the 3' sequence is targeted by dsRNA (Fig. 1). Transitive silencing was interpreted as an RdRP activity using siRNAs as primers and synthesizing a complementary RNA towards the

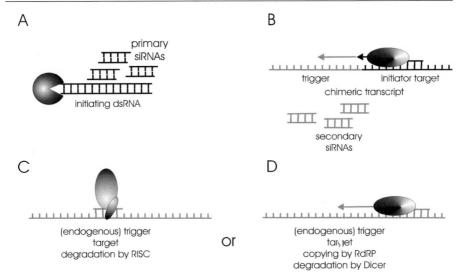

Fig. 1 Transitive silencing. A chimeric gene (**b**) is introduced into a cell. It consists of an initiator target sequence (e.g., green fluorescent protein) and a trigger sequence (e.g., from an endogenous gene). Then an initiating double-stranded RNA (*dsRNA*) (**a**) with sequences of the initiator target is introduced. The dsRNA is diced and generates primary small interfering RNAs (*siRNAs*). These serve as primers for an RNA-directed RNA polymerase (*RdRP*) on the initiator target and will be extended into the trigger sequence. The double strand is diced and generates secondary siRNAs with complementarity to the trigger target (e.g., an endogenous messenger RNA, *mRNA*). The trigger target is either sliced by an RNA-induced silencing complex (*RISC*) (**c**) or copied by RdRP and subsequently diced (**d**)

5′ end of the chimeric construct. The dsRNA produced by the RdRP would thus extend into new sequences and dicing would generate secondary siRNAs against a new target that contained these sequences (Sijen et al. 2001). These observations led to the assumption that the general RdRP function in RNAi could be the copying of the target RNA using siRNA primers, thus generating new dsRNA. This was indirectly supported by the finding that no siRNA could be detected in cells that had no target for an introduced dsRNA (panhandle RNA). Apparently, the original siRNAs generated from the panhandle RNA were rare and undetectable. Only when a target mRNA was present, secondary siRNAs, most likely derived from an RdRP-catalyzed copy on the message, were accumulated (Martens et al. 2002).

Experiments on transitive silencing revealed an unexpected feature of RdRPs: the enzyme could apparently copy only sequences in close proximity to the primary siRNA target. A spacer of more than 100 nucleotides between the primary and the secondary targets abolished transitive silencing (Alder et al. 2003; Van Houdt et al. 2003). This is difficult to understand because downstream (with respect to the RdRP) dsRNA and the corresponding siRNAs should accumulate rather than upstream dsRNAs. Even a poor processivity of RdRPs could only partially explain the observation. According to the model in Fig. 2 (see later), RdRPs should be immediately loaded with the secondary siRNAs and initiate dsRNA synthesis further and further downstream of the template.

For the recombinant *Neurospora* RdRP it has been demonstrated that it can synthesize long RNAs and short, approximately 21-nucleotide oligonucleotides on a template. Primer-dependent as well as primer-independent syntheses have been observed in vitro (Makeyev and Bamford 2002), similar to previous results with the purified tomato enzyme (Schiebel et al. 1993b). Little is known on the organization of RdRPs within the cell, for example, what proteins they are associated with and where putative complexes containing RdRPs are located.

The structure of the *Dictyostelium* RdRP and Dicer genes suggests, however, an interesting interaction. In contrast to all other Dicers identified so far , the *Dictyostelium* genes do not encode an RNA helicase domain. Instead, a Dicer-related helicase is encoded in the N-terminal part of the RdRPs. If these two proteins form a complex that requires a helicase it is probably not important which component contributes the activity. The helicase in the RdRPs and the dsRBD in the Dicer proteins may co-operate to pick up siRNAs, to unwind them and to incorporate them into the complex for primer-dependent RdRP activity. Furthermore, an RdRP–Dicer complex would directly couple dsRNA synthesis and dicing as suggested in the model

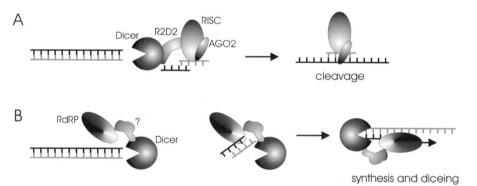

Fig. 2 RISC and RdRP as alternative partners for Dicer? Dicer associates via R2D2 with the RISC so that one strand of the newly generated siRNA can be directly transferred to the complex. After annealing with the target RNA, AGO2 cleaves only the target strand opposite the center of the siRNA (**a**). In organisms that contain RdRPs, the polymerase may directly couple to Dicer or by mediation of an unknown bridge factor. The helicase that is either part of the Dicer or the RdRP molecule may separate the siRNA strands and transfer one of them to the RdRP. A target mRNA is copied by the RdRP using the siRNA as a primer. If the complex of Dicer and RdRP is stable, the newly formed double strand may be immediately diced by the associated nuclease (**b**)

(Fig. 2). At least theoretically, a RISC would not be required in organisms that use an RdRP to convert mRNA into dsRNA and thus degrade mRNAs by Dicer. Interestingly, plants, *Dictyostelium* and others appear to have in their genomes no genes encoded that have strong similarities to R2D2, the protein that links Dicer to the RISC. Even the *C. elegans* protein RDE-4, which is supposed to be an R2D2 homolog, displays only 21% identity. On the other hand, no RdRPs appear to be encoded in the genomes of *Drosophila* and mammals and, as expected, transitive silencing could not be observed (Roignant et al. 2003). It is therefore possible that Dicer couples to RdRPs in some organisms and to the RISC in others (see later; Fig. 2) and that the function of RNAi has mechanistically diverged during evolution. In mammals, this may be explained by the PKR response, which is highly sensitive to dsRNA. The reason why *Drosophila* developed a different system remains, however, enigmatic. It is also surprising that except for the RdRPs no other obvious differences in the components of the RNAi pathway have been detected between *Drosophila* and mammals, on one hand, and RdRP containing organisms, on the other.

2.2
Multicopy Enhancers of RNAi

The RNAi components just described were mostly identified by gene knockouts or mutations that resulted in a loss of RNAi function. A different approach was used by Arndt and Raponi in *S. pombe* (Raponi and Arndt 2002, 2003). They overexpressed a complementary DNA library in a strain containing a moderately efficient antisense RNA construct against the β-galactosidase gene. They then screened for increased silencing and identified so-called antisense-enhancing sequence (aes) genes. Among several genes they identified a putative RNA helicase ded-1 and the translation–elongation factor EF-Tu as enhancers. At least for EF-Tu they could show that it increased the efficacy of RNAi as well as of antisense constructs.

Except for the *Dictyostelium* RdRPs and the putative helicase HelF (see later), no other direct comparisons have been performed in antisense-mediated and in RNAi-mediated gene silencing to our knowledge.

2.3
Inhibitors of Silencing

Most experiments on RNAi were aimed at the definition of components required for the interference mechanism. This led to the discovery of important key players as detailed before. Endogenous inhibitors of RNAi were mostly discovered fortuitously: Ruvkun's group detected eri-1, an RNase-III-related protein that is apparently responsible for the turnover of siRNAs in the cytoplasm. A mutation in eri-1 resulted in an enhancement of silencing and efficient targeting of neuronal genes that usually were refractile to RNAi

(Kennedy et al. 2004). Interestingly, eri-1 is predominantly expressed in neuronal cells and in developing gonads, suggesting a tissue-specific regulation of RNAi efficiency. The existence of eri-1 homologs has also been detected in plants (Tabler, personal communication) and in *Dictyostelium* (our own unpublished observations), but their function and localization have not yet been determined. The quantitative effect in eri-1 mutants supports the assumption that siRNAs have to accumulate above certain threshold levels in order to be efficient. At least for some genes or some tissues, a sufficient amount of siRNAs to mediate gene silencing can only be provided when degradation of the small RNAs is reduced.

Plasterk's group identified RRF-3, a putative RdRP in *C. elegans*, which, when knocked out, resulted in an enhancement of silencing. Similar to eri-1 mutants, RRF-3 knockouts allowed for improved RNAi-mediated silencing of genes in the nervous system (Simmer et al. 2002). Though the mutant strain proved very useful for practical purposes, the role of RRF-3 in the RNAi mechanism is still elusive.

In *Dictyostelium*, a putative RNA helicase HelF (unpublished data) was found by a database search for homologs of Dicer-related helicase domains. The protein contains a domain of the DEXH/D family and is localized in distinct nuclear foci. In HelF knockout strains, RNAi was considerably enhanced and some target genes that were refractile to RNAi in the wild type could be efficiently silenced in the mutant. Moreover, a strain that was inefficiently silenced by an RNAi construct displayed a very strong knockdown phenotype when the HelF gene was subsequently disrupted. This suggests that a dsRNA may exist in a cell but is targeted by HelF to make it unavailable for the RNAi pathway. HelF may recognize structural features and inactivate suboptimal dsRNAs. This may indeed be the case since relatively low amounts of dsRNA transcripts become efficient silencers in the HelF mutant. Alternatively, HelF may target all molecules with double-stranded structures with high affinity, but the capacity of the protein would be exhausted when excess amounts of the RNA are present. It would thus act as a surveillance system to avoid initiation of the RNAi cascade by low abundance and fortuitous dsRNA formation. A knockout of HelF would abolish such a surveillance system and even suboptimal dsRNA could be processed. The exact molecular function of HelF in the interference mechanism remains unknown. If the protein has RNA helicase activity it might serve to protect, for example, endogenous RNAs with extended secondary structure from being targeted by the RNAi system. However, a recent analysis revealed that at least in *Dictyostelium*, mRNAs are apparently strongly selected against containing inverted repeats that could form internal hairpin structures (Graf et al. 2004). The vast majority of so-called ATP-dependent RNA helicases have not been examined for enzymatic activity. It is therefore not clear if HelF can really unwind dsRNA. Recently, it was shown that members of the DEXH/D helicase family can replace proteins on dsRNA without unwinding the helix (Fairman et al. 2004).

In this respect, HelF could serve a function by exchanging proteins in RNAi complexes on the double-stranded molecules and thus modulate the RNAi machinery.

The effects of both eri-1 and HelF mutations support the idea that initial siRNAs quantities play a substantial role in silencing even though they are eventually amplified. If siRNAs are directly transferred from Dicer to the RISC or, as suggested in Fig. 2, from Dicer to an RdRP it is difficult to envision where HelF and eri-1 may interfere. It would be interesting to see if mutations in the inhibitors increase the extent of transitive silencing (Fig. 1). If the range of template copying by RdRPs is limited by the concentration of downstream siRNAs at least an eri-1 mutant should significantly enhance transitive silencing and allow for a longer spacer between the primary and the secondary targets.

Apart from the cellular machinery, plant viruses have evolved several suppressors of RNAi in order to circumvent by different means the plant defense system against molecular parasites. These proteins impair the RNAi pathway and thus support virus propagation. Some viral suppressors such as HC-Pro, P21 and P19 bind siRNAs and make them unavailable for the silencing mechanism (Dunoyer et al. 2004) while others, like TCV CP, may interact with Dicer and inhibit its function (Qu et al. 2003).

3
Antisense RNA Mediated Gene Silencing

Surprisingly, there is only little published work on efforts to elucidate the mechanisms of antisense RNA. Even now, there is not even clear agreement of whether gene silencing takes place in the nucleus or in the cytoplasm. Some early reports described the inhibition of splicing by natural (Munroe 1988; Munroe and Lazar 1991) and artificial antisense RNA (Volloch et al. 1991) and thus located the mechanism in the nucleus. In fact, antisense-mediated inhibition of precursor mRNA processing has been described for several experimental systems. In contrast, others found that intron sequences were less or not at all efficient as targets and thus suggested that interaction occurred in the cytoplasm. This was in agreement with the observation that mature cytoplasmic mRNAs could be abolished by the induced expression of an antisense transcript (Hildebrandt and Nellen 1992). Similarly, the mode of action is discussed controversially. In most cases, antisense RNA and the corresponding mRNA are both not detected at all on Northern blots, while their strong expression could be shown by run-on transcription experiments (Crowley et al. 1985). The general conclusion from these data was that the hybrid of sense and antisense RNA was enzymatically degraded. It was, however, rarely discussed that this would require exactly stoichiometric amounts of sense and antisense RNA in order not to find some excess sense or antisense transcripts.

Given the fact that antisense transcripts were frequently derived from multi-copy insertions of gene constructs driven by strong promoters, the antisense transcript would probably be present in vast excess. Consequently, residual antisense RNA should be detected in silenced cells. Even though this was seen in some experiments, it seems not to be the rule.

Earlier suggestions that antisense RNA may interfere with translation were not strongly pursued but have more recently regained importance in respect to microRNAs, close relatives of siRNAs that interfere with the translation machinery instead of cleaving the target RNA.

Another enigma was the observation that some genes were resistant to transgenic antisense constructs: the respective antisense RNA was expressed, sometimes in large amounts, but there was no effect on the corresponding mRNA or the encoded gene product. As discussed earlier (Nellen and Lichtenstein 1993), this strongly argued against a obligatory hybridization of RNAs in a cell since these hybrids should be immediately degraded by Dicer (Bernstein et al. 2001; Novotny et al. 2001). An alternative explanation was that the target RNA was, by folding and/or decoration with proteins, not accessible for hybridization with the antisense counterpart. By using many different oligonucleotides directed against an mRNA, a correlation between predicted secondary structure and the efficacy of oligonucleotide-mediated RNaseH cleavage was shown (Sczakiel and Far 2002). It is, however, rather unlikely that similar rules apply for the hybridization of long sense and antisense transcripts.

In some cases, different cellular localization of the complementary transcripts may explain the failure to form diceable double strands and thus the failure to induce silencing. When antisense transcripts are microinjected into cells or organisms, cellular localization should, at least initially, play a minor role. Nevertheless, some silencing experiments failed and thus supported the assumption that in vivo hybridization was not an obligatory event.

Fire et al. (1998) proposed that antisense RNA mediated gene silencing was possibly due to contaminating sense strands in microinjection experiments. These would thus generate small amounts of dsRNA. Similarly, it had previously been proposed that, especially in transfected plants, spurious transcription from a cryptic promoter in the opposite orientation could generate complementary RNA and lead to dsRNA formation. In fact, this assumption was used to explain the enigmatic "co-suppression" phenomenon, where the introduction of a sense transgene resulted in the silencing of both the endo-gene and the transgene. The problem was, however, only shifted to a different level: how could substoichiometric amounts of an antisense RNA result in complete gene silencing? In some reports it was suggested that an antisense transcript was cleaved into small fragments that could each target and destroy a different mRNA (Flavell 1994). This is reminiscent of the RNAi pathway where many siRNAs could be generated from a dsRNA and a single siRNA on the RISC could mediate the degradation of a transcript.

4
Antisense RNA and RNAi

With the discovery of key players in RNAi mechanisms it was possible to address the question of the putative identity or similarity between RNAi and antisense RNA on a solid basis by asking if the same cellular components are required and the same mechanisms are applied for both systems. It is interesting to note in this context that antisense-mediated silencing could never be achieved in *Saccharomyces cerevisiae* (Atkins et al. 1994). Similarly there are no reports of successful RNAi experiments in baker's yeast. This is not surprising since *S. cerevisiae* appears to be one of very few eukaryots that do not encode essential proteins of the RNAi pathway in their genome (Sigova et al. 2004). Nevertheless, the coincidence that both mechanisms are not functional may argue that they are identical or closely related.

4.1
Small Interfering RNAs

Only in *S. pombe* (Raponi and Arndt 2002, 2003) and *Dictyostelium* (our unpublished data) has it been shown unambiguously that antisense RNA and RNAi constructs both result in the generation of siRNAs. This observation could be explained either by hybridization between mRNA and antisense RNA or by conversion of the antisense transcript to a dsRNA by a secondary reaction. In both cases, targets for Dicer would be generated. The data clearly showed that Dicer was involved in both pathways but did not rule out differences. The results are also consistent with the suggestion that the mechanisms are identical and that the antisense experiments were only functional because of dsRNA formation. An observation had, however, cast some doubt on this assumption: using the same target for antisense- and RNAi-mediated silencing higher quantities of siRNAs in the antisense transformants than in the RNAi (inverted repeat) transformants were observed (Martens and Nellen, unpublished results). In contrast, the accumulation of siRNAs in *S. pombe* antisense transformants was significantly lower compared with that in RNAi experiments (Raponi and Arndt 2003). This discrepancy could be due to different silencing levels in the respective strains and deserves further investigation.

4.2
RNA-Directed RNA Polymerases

In *Dictyostelium*, three RdRPs, rrpA, rrpB and rrpC, are encoded in the genome. While rrpA and rrpB differ by only 4% in the encoded amino acid sequence, rrpC is more divergent (Martens et al. 2002). Knockouts of rrpA and rrpB have been generated and they displayed no obvious mutant phenotype. In contrast, the knockout of rrpC is impaired in development. When the three gene disruption strains were examined for the efficiency of antisense RNA, they all failed to display silencing. In comparison with the wild type, antisense RNA was accumulated in these cells to different levels but had apparently no effect on the mRNA stability. In contrast, RNAi was only impaired in rrpA mutants but was as efficient as in the wild type in the rrpB and rrpC minus strains. This observation indicated that the three enzymes had distinct functions in RNAi and antisense pathways and that consequently, the silencing mechanisms differed at least in their requirements for RdRPs.

Since antisense RNA and RNAi apparently have partially overlapping but distinct pathways, it was interesting to see if co-expression of both silencing constructs had an additive or even a synergistic effect. In fact, co-transfection experiments did not only add up in effect but rather increased silencing synergistically. Taken together, the data strongly suggest that the two mechanisms are related but not identical.

4.3
HelF

Further supporting evidence for the difference between antisense RNA and RNAi emerged from experiments with the *Dictyostelium* HelF disruption strain. These cells that displayed significantly enhanced silencing by dsRNA for all genes tested did not show any increased sensitivity to antisense RNA. HelF is predominantly localized in the nucleus and may therefore regulate an early, nuclear step in the RNAi pathway. In contrast, the antisense mechanism, which is presumably mediated by cytoplasmic RdRPs, remains unaffected.

5
Models

The following model (Fig. 3) combines the current knowledge on RNAi and antisense mediated gene silencing. Some parts of the suggested pathway are speculations that are, however, consistent with the available data.

The initial dsRNA is diced to siRNAs. According to the standard model of RNAi, siRNAs could be transferred to the RISC and mediate slicing of the target mRNA. Alternatively or in a parallel pathway, the antisense strand of the siRNA could be transferred to an RdRP and serve as a primer on the target transcript. The newly generated dsRNA would provide a target for Dicer, the mRNA would be destroyed and more siRNAs would be produced. As mentioned before, it is clear that the majority of the siRNAs are not generated from the original double strand but rather are derived from the target, most likely by means of an RdRP activity and subsequent dicing (Martens et al. 2002).

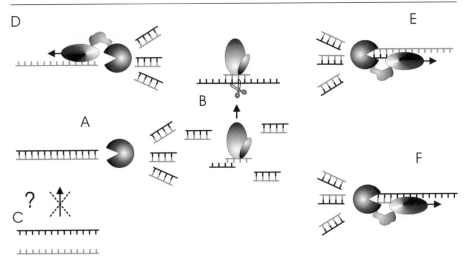

Fig. 3 Hypothetical pathway for RNA interference (*RNAi*) and antisense-mediated gene silencing. dsRNA is degraded to 21mer siRNAs by Dicer (**a**). The siRNAs are unwound by RISC/R2D2 (**b**) and transferred to the mRNA (*black*) that is cut by the Slicer activity in RISC. Antisense RNA (*gray*) and mRNA (*black*) most likely do not spontaneously form hybrids (**c**). Possibly, a primer-independent RdRP recognizes aberrant antisense RNA and copies it. The RdRP may be directly linked to Dicer; thus, dsRNA synthesis and dicing would be coupled (**d**). The siRNA produced in **d** may be picked up by RISC (**b**) or by a primer-dependent RdRP. Since the sense or the antisense strand of the siRNA may be transferred, this may target either the antisense RNA (**e**) that has not yet been copied by the primer-independent RdRP (**d**) or the mRNA (**f**). The primed RdRP may also be associated with Dicer, which immediately processes the double strand to siRNAs. This would result in degradation of both RNA strands independent of their respective concentration and result in a higher concentration of siRNAs. The RNAi and the antisense pathway may have threshold levels for diceable dsRNA and aberrant RdRP targets, respectively, to produce siRNAs. Co-expression of both antisense RNA and RNAi provides two sources for siRNA that add up and may exceed the critical concentration to initiate the silencing cascade. Note that formally, a RISC is not required in organisms containing RdRPs or a putative Dicer/RdRP complex

It is interesting to note that the concerted action of an RdRP and Dicer could be sufficient for mRNA degradation and, at least in theory, RISC would not be required.

In the case of antisense RNA, the direct hybridization between the complementary strands is unlikely as explained earlier.

As has been discussed extensively for plants, some RNAs, especially antisense RNAs and some transcripts from transgenes, could be recognized by the cell as "aberrant" (Flavell 1994) and function as inducers of silencing. Aberrant RNAs could be specified by incorrect processing and/or association with specific proteins. In *Arabidopsis* it has been suggested that uncapped mRNAs become exposed to RdRPs and may initiate RNAi (Gazzani et al. 2004). In

C. elegans, silencing of a green fluorescent protein (GFP) transgene could be achieved by an approximately 85 nucleotide RNA isolated from silenced GFP plants (Boutla et al. 2002). The authors suggested that this RNA that could not be further specified could represent an aberrant silencing inducer.

Even though aberrant RNAs still await a precise definition, they are apparently recognized by the cell and serve as a signal to initiate RNAi or a related mechanism.

In contrast to the primer-dependent activity suggested earlier, aberrant RNAs may be copied by RdRPs in a primer-independent way. This could explain the requirement for at least two different enzymes in the antisense mechanism in *Dictyostelium*: a primer-dependent and a primer-independent one. The requirements for different RdRPs in RNAi and antisense mechanisms suggest that these activities may be carried out by distinct enzymes. dsRNA synthesized by a primer-independent RdRP on aberrant transcripts would be diced and generate siRNAs. These may become associated with RISC and/or serve as primers for a primer-dependent RdRP. Possibly, aberrancy of an RNA is not defined as an absolute feature. Depending on folding or decoration with proteins, some molecules within, for example, an antisense RNA population may be strong targets and others may be weak targets for the primer-independent RdRP. The mechanism described would first copy the most aberrant RNAs and convert them to Dicer substrates. The resulting siRNAs could then serve as primers for a different RdRP that copies the mRNA and those antisense molecules that were not sufficiently aberrant to be recognized by the primer-independent RdRP. It should be noted that in the case of antisense RNA, the cell would contain two targets for siRNA-mediated degradation, the endogenous mRNA and the antisense transgene transcript. Both would be degraded independently of their relative concentrations and, consistent with the observations, more siRNAs would be accumulated.

The synergistic effect of antisense RNA and RNAi could also be explained by the model. Antisense RNAs that are weak targets for the primer-independent RdRPs may generate siRNAs below a threshold level. With the co-expression of RNAi, a sufficient number of siRNAs would be provided to induce the chain reaction by siRNA production via the primer-dependent RdRP. Similarly, some dsRNAs may be poor targets for Dicer and co-expression of an antisense RNA could raise the amount of siRNAs above the threshold level.

The biological functions and the different effects on gene silencing of the putative RNA helicase HelF in *Dictyostelium* are still elusive. HelF is predominantly found in the nucleus and may decrease RNAi efficiency by different mechanisms: the protein could unwind fold-back structures in the nucleus and thus lead to nuclear degradation of an RNAi effector or to the inhibition of export to the cytoplasm. Alternatively, HelF may compete with a putative nuclear Dicer and thus decrease siRNA formation. In fact, we have observed that HelF inhibits Dicer activity in vitro. Even though the generation of siRNAs is presumably mostly cytoplasmic, nuclear dicing cannot be excluded. As already suggested, antisense RNA would be copied by RdRPs in the cytoplasm and thus evade the nuclear helicase.

The effect of eri-1 on antisense RNA has not yet been examined. Since siRNAs are generated by antisense RNA and are crucial components for the degradation of both target transcripts, one would predict that eri-1 knockouts should enhance antisense RNA as well as RNAi.

Taken together, there is accumulating evidence that antisense-mediated gene silencing and RNAi use overlapping but partially different pathways. The model proposed in this paper attributes a crucial role to RdRPs in antisense mechanisms. This indirectly suggests that the action of antisense may be fundamentally different in organisms that do not encode RdRPs.

Acknowledgements The work was supported by a grant from the Deutsche Forschungsgemeinschaft (Ne285/8, SPP1129) and from the EU (FP6, 5120).

References

Alder MN, Dames S, Gaudet J, Mango SE (2003) Gene silencing in Caenorhabditis elegans by transitive RNA interference. RNA 9:25–32

Atkins D, Arndt GM, Izant JG (1994) Antisense gene expression in yeast. Biol Chem Hoppe Seyler 375:721–729

Bass BL (2002) RNA editing by adenosine deaminases that act on RNA. Annu Rev Biochem 71:817–846

Bernstein E, Caudy AA, Hammond SM, Hannon GJ (2001) Role for a bidentate ribonuclease in the initiation step of RNA interference. Nature 409:363–366

Billy E, Brondani V, Zhang H, Muller U, Filipowicz W (2001) Specific interference with gene expression induced by long, double-stranded RNA in mouse embryonal teratocarcinoma cell lines. Proc Natl Acad Sci USA 98:14428–14433

Blaszczyk J, Tropea JE, Bubunenko M, Routzahn KM, Waugh DS, Court DL, Ji X (2001) Crystallographic and modeling studies of RNase III suggest a mechanism for double-stranded RNA cleavage. Structure 9:1225–1236

Boutla A, Kalantidis K, Tavernarakis N, Tsagris M, Tabler M (2002) Induction of RNA interference in Caenorhabditis elegans by RNAs derived from plants exhibiting post-transcriptional gene silencing. Nucleic Acids Res 30:1688–1694

Brantl S (2002) Antisense-RNA regulation and RNA interference. Biochim Biophys Acta 1575:15–25

Cogoni C, Macino G (1999) Gene silencing in Neurospora crassa requires a protein homologous to RNA-dependent RNA polymerase. Nature 399:166–169

Coleman J, Hirashima A, Inokuchi Y, Green PJ, Inouye M (1985) A novel immune system against bacteriophage infection using complementary RNA (micRNA). Nature 315:601–603

Crowley TE, Nellen W, Gomer RH, Firtel RA (1985) Phenocopy of discoidin I-minus mutants by antisense transformation in Dictyostelium. Cell 43:633–641

Dalmay T, Hamilton A, Rudd S, Angell S, Baulcombe DC (2000) An RNA-dependent RNA polymerase gene in Arabidopsis is required for posttranscriptional gene silencing mediated by a transgene but not by a virus. Cell 101:543–553

Dunoyer P, Lecellier CH, Parizotto EA, Himber C, Voinnet O (2004) Probing the microRNA and small interfering RNA pathways with virus-encoded suppressors of RNA silencing. Plant Cell 16:1235–1250

Elbashir SM, Harborth J, Lendeckel W, Yalcin A, Weber K, Tuschl T (2001) Duplexes of 21-nucleotide RNAs mediate RNA interference in cultured mammalian cells. Nature 411:494–498

Fairman ME, Maroney PA, Wang W, Bowers HA, Gollnick P, Nilsen TW, Jankowsky E (2004) Protein displacement by DExH/D "RNA helicases" without duplex unwinding. Science 304:730–734

Fire A, Xu S, Montgomery MK, Kostas SA, Driver, SE, Mello CC (1998) Potent and specific genetic interference by double-stranded RNA in Caenorhabditis elegans. Nature 391:806–811

Flavell RB (1994) Inactivation of gene expression in plants as a consequence of specific sequence duplication. Proc Natl Acad Sci USA 91:3490–3496

Gazzani S, Lawrenson T, Woodward C, Headon D, Sablowski R (2004) A link between mRNA turnover and RNA interference in Arabidopsis. Science 306:1046–1048

Graf S, Borisova BE, Nellen W, Steger G, Hammann C (2004) A database search for double-strand containing RNAs in Dictyostelium discoideum. Biol Chem 385:961–965

Hall IM, Shankaranarayana GD, Noma K, Ayoub N, Cohen A, Grewal SI (2002) Establishment and maintenance of a heterochromatin domain. Science 297:2232–2237

Hammond SM, Bernstein E, Beach D, Hannon GJ (2000) An RNA-directed nuclease mediates post-transcriptional gene silencing in Drosophila cells. Nature 404:293–296

Hammond SM, Boettcher S, Caudy AA, Kobayashi R, Hannon GJ (2001) Argonaute 2, a link between genetic and biochemical analyses of RNAi. Science 293:1146–1150

Hildebrandt M, Nellen W (1992) Differential antisense transcription from the Dictyostelium EB4 gene locus: implications on antisense-mediated regulation of mRNA stability. Cell 69:197–204

Kennedy S, Wang D, Ruvkun G (2004) A conserved siRNA-degrading RNase negatively regulates RNA interference in C. elegans. Nature 427:645–649

Khvorova A, Reynolds A, Jayasena SD (2003) Functional siRNAs and miRNAs exhibit strand bias. Cell 115:209–216

Lee RC, Feinbaum RL, Ambros V (1993) The C. elegans heterochronic gene lin-4 encodes small RNAs with antisense complementarity to lin-14. Cell 75:843–854

Lee YS, Nakahara K, Pham JW, Kim K, He Z, Sontheimer EJ, Carthew RW (2004) Distinct roles for Drosophila Dicer-1 and Dicer-2 in the siRNA/miRNA silencing pathways. Cell 117:69–81

Liu J, Carmell MA, Rivas FV, Marsden CG, Thomson JM, Song JJ, Hammond SM, Joshua-Tor L, Hannon GJ (2004) Argonaute 2 is the catalytic engine of mammalian RNAi. Science 305:1437–1441

Liu Q, Rand TA, Kalidas S, Du F, Kim HE, Smith DP, Wang X (2003) R2D2, a bridge between the initiation and effector steps of the Drosophila RNAi pathway. Science 301:1921–1925

Makeyev EV, Bamford DH (2002) Cellular RNA-dependent RNA polymerase involved in posttranscriptional gene silencing has two distinct activity modes. Mol Cell 10:1417–1427

Martens H, Novotny J, Oberstrass J, Steck TL, Postlethwait P, Nellen W (2002) RNAi in Dictyostelium: the role of RNA-directed RNA polymerases and double-stranded RNase. Mol Biol Cell 13:445–453

Meister G, Landthaler M, Patkaniowska A, Dorsett Y, Teng G, Tuschl T (2004) Human Argonaute 2 mediates RNA cleavage targeted by miRNAs and siRNAs. Mol Cell 15:185–197

Mizuno T, Chou MY, Inouye M (1984) A unique mechanism regulating gene expression: translational inhibition by a complementary RNA transcript (micRNA). Proc Natl Acad Sci US A 81:1966–1970

Möhrle A (1989) Nachweis, Anreinigung und Charakterisierung einer doppelstrangspezifischen Ribonuklease in *Dictyostelium discoideum*. Diploma thesis, Ludwig-Maximilian-Universität, Munich, Germany

Munroe SH (1988) Antisense RNA inhibits splicing of pre-mRNA in vitro. EMBO J 7:2523–2532

Munroe SH, Lazar MA (1991) Inhibition of c-erbA mRNA splicing by a naturally occurring antisense RNA. J Biol Chem 266:22083–22086

Nellen W, Lichtenstein C (1993) What makes an mRNA anti-sense-itive? Trends Biochem Sci 18:419–423

Novotny J, Diegel S, Schirmacher H, Mohrle A, Hildebrandt M, Oberstrass J, Nellen W (2001) Dictyostelium double-stranded ribonuclease. Methods Enzymol 342:193–212

Orban TI, Izaurralde E (2005) Decay of mRNAs targeted by RISC requires XRN1, the Ski complex, and the exosome. RNA 11:459–469

Persengiev SP, Zhu X, Green MR (2004) Nonspecific, concentration-dependent stimulation and repression of mammalian gene expression by small interfering RNAs (siRNAs). RNA 10:12–18

Pestka S, Langer JA, Zoon KC, Samuel CE (1987) Interferons and their actions. Annu Rev Biochem 56:727–777

Pham JW, Pellino JL, Lee YS, Carthew RW, Sontheimer EJ (2004) A Dicer-2-dependent 80s complex cleaves targeted mRNAs during RNAi in Drosophila. Cell 117:83–94

Qu F, Ren T, Morris TJ (2003) The coat protein of turnip crinkle virus suppresses post-transcriptional gene silencing at an early initiation step. J Virol 77:511–522

Raponi M, Arndt GM (2002) Dominant genetic screen for cofactors that enhance antisense RNA-mediated gene silencing in fission yeast. Nucleic Acids Res 30:2546–2554

Raponi M, Arndt GM (2003) Double-stranded RNA-mediated gene silencing in fission yeast. Nucleic Acids Res 31:4481–4489

Roignant JY, Carre C, Mugat B, Szymczak D, Lepesant JA, Antoniewski C (2003) Absence of transitive and systemic pathways allows cell-specific and isoform-specific RNAi in Drosophila. RNA 9:299–308

Schiebel W, Haas B, Marinkovic S, Klanner A, Sanger HL (1993a) RNA-directed RNA polymerase from tomato leaves. I. Purification and physical properties. J Biol Chem 268:11851–11857

Schiebel W, Haas B, Marinkovic S, Klanner A, Sanger HL (1993b) RNA-directed RNA polymerase from tomato leaves. II. Catalytic in vitro properties. J Biol Chem 268:11858–11867

Schmedt C, Green SR, Manche L, Taylor DR, Ma Y, Mathews MB (1995) Functional characterization of the RNA-binding domain and motif of the double-stranded RNA-dependent protein kinase DAI (PKR). J Mol Biol 249:29–44

Schwarz DS, Hutvagner G, Du T, Xu Z, Aronin N, Zamore PD (2003) Asymmetry in the assembly of the RNAi enzyme complex. Cell 115:199–208

Sczakiel G, Far RK (2002) The role of target accessibility for antisense inhibition. Curr Opin Mol Ther 4:149–153

Sigova A, Rhind N, Zamore PD (2004) A single Argonaute protein mediates both transcriptional and posttranscriptional silencing in Schizosaccharomyces pombe. Genes Dev 18:2359–2367

Sijen T, Fleenor J, Simmer F, Thijssen KL, Parrish S, Timmons L, Plasterk RH, Fire A (2001) On the role of RNA amplification in dsRNA-triggered gene silencing. Cell 107:465–476

Simmer F, Tijsterman M, Parrish S, Koushika SP, Nonet ML, Fire A, Ahringer J, Plasterk RH (2002) Loss of the putative RNA-directed RNA polymerase RRF-3 makes C. elegans hypersensitive to RNAi. Curr Biol 12:1317–1319

Sledz CA, Holko M, de Veer MJ, Silverman RH, Williams BR (2003) Activation of the interferon system by short-interfering RNAs. Nat Cell Biol 5:834–839

Smardon A, Spoerke JM, Stacey SC, Klein ME, Mackin N, Maine EM (2000) EGO-1 is related to RNA-directed RNA polymerase and functions in germ-line development and RNA interference in C. elegans. Curr Biol 10:169–178

Song JJ, Smith SK, Hannon GJ, Joshua-Tor L (2004) Crystal structure of Argonaute and its implications for RISC slicer activity. Science 305:1434–1437

Tomari Y, Matranga C, Haley B, Martinez N, Zamore PD (2004) A protein sensor for siRNA asymmetry. Science 306:1377–1380

Tonkin LA, Bass BL (2003) Mutations in RNAi rescue aberrant chemotaxis of ADAR mutants. Science 302:1725

Van Houdt H, Bleys A, Depicker A (2003) RNA target sequences promote spreading of RNA silencing. Plant Physiol 131:245–253

Volloch V, Schweitzer B, Rits S (1991) Inhibition of pre-mRNA splicing by antisense RNA in vitro: effect of RNA containing sequences complementary to exons. Biochem Biophys Res Commun 179:1593–1599

Wightman B, Ha I, Ruvkun G (1993) Posttranscriptional regulation of the heterochronic gene lin-14 by lin-4 mediates temporal pattern formation in C. elegans. Cell 75:855–862

Zhang H, Kolb FA, Jaskiewicz L, Westhof E, Filipowicz W (2004) Single processing center models for human Dicer and bacterial RNase III. Cell 118:57–68

Nucleic Acids and Molecular Biology, Vol. 17
Wolfgang Nellen, Christian Hammann (Eds.)
Small RNAs
© Springer-Verlag Berlin Heidelberg 2005

Epigenetic Silencing of Transposons in the Green Alga *Chlamydomonas reinhardtii*

Karin van Dijk · Hengping Xu · Heriberto Cerutti (✉)

School of Biological Sciences and Plant Science Initiative,
University of Nebraska-Lincoln, E211 Beadle Center, P.O. Box 880666,
Lincoln, NE 68588-0666, USA
hcerutti1@unl.edu

Abstract Transposons are mobile genetic elements that live parasitically within the genome of cellular organisms. They can affect the fitness of their hosts by influencing gene function, gene activity, genome structure, and overall DNA content. Since excessive transposon activity can result in a high mutagenic rate and genomic instability, eukaryotes have evolved epigenetic mechanisms to reduce transposition to manageable levels. The alga *Chlamydomonas reinhardtii* appears to have several, at least partly independent, transposon repression pathways that operate at either the transcriptional or the post-transcriptional level. Two genes have been implicated in the transcriptional silencing of transposons and single-copy transgenes: *Mut9*, which encodes a novel serine/threonine protein kinase capable of phosphorylating histones H3 and H2A, and *Mut11*, which encodes a WD40-repeat containing protein. The Mut11 protein functions as a subunit of a histone methyltransferase complex(es) that is required for monomethylation of histone H3 lysine 4 and the maintenance of repressed euchromatic domains. These mechanisms of transcriptional gene silencing operate independently from the RNA interference (RNAi) machinery. In contrast, a putative DEAH-box RNA helicase (Mut6) has been demonstrated to participate in the post-transcriptional suppression of transgenes and transposons. The Mut6 helicase is homologous to an essential *Saccharomyces cerevisiae* precursor messenger RNA (pre-mRNA) splicing factor and appears to be required for the processing of specific *Chlamydomonas* pre-mRNAs, including those corresponding to components of the RNAi machinery such as Dicer and Ago1. These findings implicate RNAi in the post-transcriptional repression of transposons in *Chlamydomonas*. Thus, multiple silencing mechanisms appear to operate as a defense system against the expansion of transposable elements in algae. These independent repression pathways likely increase the genetic robustness of transposon silencing against mutational and environmental perturbations.

1
Introduction

Transposons are genetic elements that can move around to different positions within a genome. Since the discovery of these "jumping genes" by McClintok (1984) in maize, transposable elements have been identified in many organisms and are generally classified into two classes on the basis of their structure and their mechanism of transposition. Class I transposons (or retrotransposons) move via an RNA intermediate that is first copied by a reverse transcriptase into DNA followed by insertion of the complementary DNA copies into new genomic sites (Okamoto and Hirochika 2001; Feschotte et al. 2002). Class II transposons (or DNA elements) move as DNA segments, either via a "cut-and-paste" mechanism (involving a transposon-encoded transposase) or via a replicative mechanism (Okamoto and Hirochika 2001; Feschotte et al. 2002; Vastenhouw and Plasterk 2004). Several transposable elements have been described in the unicellular green alga *Chlamydomonas reinhardtii*, including a class I retrotransposon, *TOC1*, and a number of class II elements, *Gulliver*, *Pioneer*, *Tcr1*, *Tcr2*, and *Tcr3* (Day et al. 1988; Ferris 1989; Schnell and Lefebvre 1993; Wang et al. 1998). With the availability of complete genome sequences, it is becoming apparent that transposable elements or their remnants make up a significant part of many eukaryotic genomes. For instance, approximately 45% of the human genome consists of (retro)transposon-derived sequences (Lander et al. 2001) and an even higher proportion of the maize genome is estimated to comprise transposon-related sequences (SanMiguel and Bennetzen 1998; Okamoto and Hirochika 2001).

Transposons are generally considered to be selfish DNA, parasites that live within the genome of cellular organisms. In this respect they are similar to viruses. In fact, some (retro)viruses and (retro)transposons show comparable features in their genome structure and encoded biochemical activities, suggesting that they have shared a common evolutionary ancestor (Okamoto and Hirochika 2001; Havecker et al. 2004). Transposable elements can affect the fitness of their hosts by changing gene function or activity (Nekrutenko and Li 2001; Lippman et al. 2004). They can also influence the structure and DNA content of a genome (Feschotte et al. 2002). Even transposon-derived duplications that are no longer capable of movement can potentially lead to genome rearrangements by recombination-based mechanisms. These effects can have an important role in the diversification and evolution of species (McClintok 1984; Nekrutenko and Li 2001; Feschotte et al. 2002) but often they are merely deleterious to individuals (Bestor 2003). For instance, in *Chlamydomonas*, active transposable elements have been identified on the basis of their ability to cause mutations disrupting gene function (Day et al. 1988; Schnell and Lefebvre 1993; Wang et al. 1998). Likewise, retrotransposon integration into biologically important genes has been implicated in several human diseases (Miki 1998; Lower 1999; Colombo et al. 2000). Since excessive transposon activity can result in a high mutagenic rate and genome instability, many organisms appear to have developed mechanisms to reduce transposition to a manageable level. In addition, packaging of transposable elements into repressive (hetero)chromatin likely prevents chromosomal rearrangements by illegitimate recombination between ectopic repeat sequences.

Epigenetic processes, which result in heritable changes in gene expression without modifications in DNA sequence, play important roles in the cellular responses to transposable elements and viruses, as well as in the control of organismal development (Matzke et al. 2001; Zamore 2002; Carrington and Ambros 2003; Cerutti 2003; Baulcombe 2004; Lippman and Martienssen 2004; Vastenhouw and Plasterk 2004). In protists, plants, fungi, and animals, epigenetic silencing mechanisms have also been implicated in the suppression of transgene expression. Depending on the level at which silencing occurs, two types of phenomena have been distinguished in photosynthetic eukaryotes: transcriptional gene silencing (TGS) and post-transcriptional gene silencing (PTGS) (Matzke et al. 2001; Vaucheret and Fagard 2001; Jeong et al. 2002). TGS entails transcriptional suppression, whereas PTGS involves a variety of RNA-mediated silencing phenomena, which may result in target RNA degradation or translational repression. However, in some organisms, RNA-mediated processes can also lead to transcriptional silencing by inducing DNA methylation and/or heterochromatin formation (Cerutti 2003; Denli and Hannon 2003; Baulcombe 2004; Meister and Tuschl 2004). Here, we summarize recent advances in our mechanistic understanding of transposon silencing in algal systems, with the main emphasis on *C. reinhardtii*.

2
Post-Transcriptional Gene Silencing

2.1
General Mechanisms

In higher plants, algae (including *Chlamydomonas*), and some fungi, the overexpression or misexpression of transgenes can induce the silencing of homologous sequences (i.e., transgene and endogenous gene sequence) at the post-transcriptional level (Jorgensen et al. 1999; Cogoni and Macino 2000; Wu-Scharf et al. 2000; Baulcombe 2004). Some, if not all, of these effects are mechanistically related to the RNA interference (RNAi) phenomenon, first described in *Caenorhabditis elegans* and now known to be widespread in many eukaryotes (Fire 1999; Zamore 2002; Cerutti 2003; Denli and Hannon 2003; Baulcombe 2004). Indeed, double-stranded RNA (dsRNA) has been demonstrated to induce the degradation of homologous RNAs in organisms as diverse as protists, plants, fungi, and animals. In some species, the RNAi machinery is also involved in the processing of microRNAs (miRNAs) and the translational repression of target messenger RNAs (mRNAs) (Carrington and Ambros 2003; Ambros 2004; Bartel 2004). Besides these post-transcriptional effects, RNA-mediated processes have also been implicated in heterochromation formation, DNA methylation, and even DNA elimination (Cerutti 2003; Meister and Tuschl 2004).

Currently, the most extensively characterized dsRNA-mediated mechanism is targeted mRNA degradation guided by small interfering RNAs (siRNAs). Genetic and biochemical studies from diverse species have revealed that long dsRNAs are processed into siRNAs of about 21–27 nucleotides by an RNase-III-like endonuclease, named Dicer (Zamore 2002; Denli and Hannon 2003; Meister and Tuschl 2004). siRNAs are then incorporated into a multiprotein complex, the RNA-induced silencing complex (RISC), via a series of intermediate subcomplexes (Pham et al. 2004; Tomari et al. 2004). During this process, the double-stranded siRNA is unwound and only a single-stranded molecule is retained in the final active complex. The activated RISC then functions as a multiple-turnover enzyme that recognizes and cleaves mRNA molecules complementary to the incorporated single-stranded siRNA (Meister and Tuschl 2004; Haley and Zamore 2004). In flies, worms, and humans, members of the Argonaute family of proteins are core components of the RISC or RNA-induced silencing like complexes (Liu et al. 2004; Meister and Tuschl 2004). Moreover, Argonaute proteins are required genetically for RNA silencing in every organism where their function has been studied (Carmell et al. 2002; Baulcombe 2004; Sigova et al. 2004). Even though plants and animals contain multiple Argonaute paralogs, among the human Argonaute proteins, only Ago2 can mediate miRNA- or siRNA-directed endonucleolytic cleavage of target RNA (Liu et al. 2004; Meister and Tuschl 2004). The recent three-dimensional structure of archaeal Argonaute proteins, together with experiments evaluating the importance of predicted catalytic residues in human Ago2, suggest that this protein itself is the RISC endoribonuclease (Liu et al. 2004; Parker et al. 2004; Song et al. 2004). In contrast, other Argonaute paralogs may have specialized to direct translational repression of mRNAs or transcriptional silencing of DNA sequences.

In several species, RNA-dependent RNA polymerases (RdRPs) also play an important role in the RNAi process. For instance, putative homologs of a tomato RdRP are required for PTGS triggered by sense transgenes in *Arabidopsis thaliana*, for quelling (a phenomenon similar to PTGS) in *Neurospora crassa*, and for RNAi in *C. elegans* and *Dictyostelium discoideum* (Cogoni and Macino 2000; Sijen et al. 2001; Martens et al. 2002; Cerutti 2003; Baulcombe 2004). It has been proposed that RdRPs generate dsRNA from single-stranded transcripts either by de novo, primer-independent second-strand synthesis (utilizing as a template "aberrant" RNAs, presumably lacking normal processing signals such as a 5′ cap or a polyadenosine tail) or by using siRNAs as primers to synthesize additional dsRNA (employing as a template the target mRNAs) (Sijen et al. 2001; Cerutti 2003; Baulcombe 2004). Thus, RdRP activity may initiate RNAi (by producing the trigger dsRNA) or dramatically enhance the RNAi response (by amplifying the amount of dsRNA to be processed into siRNAs by Dicer) (Baulcombe 2004). However, no RdRPs homologs have been identified in *Drosophila melanogaster* and mammals, suggesting that dsRNA-induced RNAi can occur in the absence of RdRP activity (Zamore 2002; Baulcombe 2004). Likewise, in the current draft version of the *C. reinhardtii* genome (http://genome.

jgi-psf.org/chlre2/chlre2.home.html) there is no recognizable RdRP homolog, whereas RNAi can be effectively induced by the expression of inverted repeat transgenes (Rohr et al. 2004).

The post-transcriptional silencing mechanism mediated by miRNAs is related to the action of siRNAs, but is much less understood. miRNAs closely resemble siRNAs in size and structure but they are encoded in the genome as stem-loop precursors (Carrington and Ambros 2003; Ambros 2004; Bartel 2004). Many miRNAs regulate gene expression by binding and inhibiting the translation of mRNAs, although in higher plants and in some instances in mammals, miRNAs can also target mRNAs for degradation (Carrington and Ambros 2003; Bartel 2004; Baulcombe 2004). The mechanistic differences between siRNAs and miRNAs have been attributed partly to the degree of complementarity between the short RNAs and their targets. It has been proposed that for the RISC to catalyze mRNA cleavage a contiguous A-form helix must be formed between the short guide RNA and the mRNA target in the region of the scissile bond and towards the 3′ end of the siRNA (Haley and Zamore 2004; Parker et al. 2004). Presumably, this is required to generate the correct reaction geometry in the RISC active site. This would then be sufficient to explain the observed mechanistic differences between perfectly complementary siRNAs and imperfectly complementary miRNAs (Hutvágner and Zamore 2002; Doench et al. 2003).

2.2
PTGS and Transposon Silencing

Evidence in several eukaryotes indicates that one mechanism of transposon silencing shares components with the RNAi machinery (Aravin et al. 2001; Sijen and Plasterk 2003; Chicas et al. 2004; Lippman and Martienssen 2004; Vastenhouw and Plasterk 2004). However, in many cases it is not clear whether RNAi-mediated transposon suppression functions transcriptionally (mediating heterochromatin formation and/or DNA methylation), post-transcriptionally (inducing RNA degradation), or both. In principle, transposons could be silenced post-transcriptionally if the RNA replication intermediate (in the case of retrotransposons) or mRNAs encoding reverse transcriptase, transposase, and/or other auxiliary proteins (Okamoto and Hirochika 2001) are targeted for siRNA-mediated degradation. It is likely that the trigger for this transposon silencing pathway is dsRNA, which could be produced in at least three different ways (Vastenhouw and Plasterk 2004): (1) in organisms containing RdRPs, aberrant single-stranded transposon transcripts could be used as a template for the synthesis of dsRNA; (2) readthrough transcription from gene promoters neighboring an inserted transposon could result in the production of sense and antisense RNAs that anneal forming dsRNA; and/or (3) in the case of transposable elements with terminal inverted repeats (TIRs), sense transcripts that span the whole element and include both TIRs could form dsRNA by snapping back into a panhandle structure. In *C. elegans*, transposon-derived dsRNA and siRNAs of the Tc1/*mariner* family were mainly detected for the TIRs, suggesting that the main source of dsRNA is the intramolecular panhandle structure (Sijen and Plasterk 2003). However, dsRNA and siRNAs derived from internal transposon sequences could also be identified in smaller amounts (Ambros et al. 2003; Sijen and Plasterk 2003). siRNAs corresponding to transposable elements have also been detected in, among other species, higher plants, *Trypanosoma brucei*, *N. crassa*, and *D. melanogaster* (Djikeng et al. 2001; Llave et al. 2002; Aravin et al. 2003; Rudenko et al. 2003; Chicas et al. 2004).

In the green alga *C. reinhardtii*, long antisense RNAs corresponding to the *TOC1* retrotransposon have been observed in some strains and postulated to arise by read-through transcription from a promoter(s) flanking a specific transposon insert(s) (Day and Rochaix 1991). Moreover, by genetic tetrad analyses, the production of sense and antisense transcripts (presumably annealing into dsRNA) in the same cell has been shown to inhibit *TOC1* RNA accumulation (Day and Rochaix 1991). Consistent with these findings, we have detected siRNAs derived from the *TOC1* long terminal repeats, likely generated by Dicer processing of dsRNA (Fig. 1). The *TOC1* siRNAs are slightly smaller than those produced in strains expressing dsRNA from inverted repeat transgenes (Fig. 1a). We have also observed small RNAs derived from the class II transposon *Gulliver*, but these were only noticeable in a mutant strain with defects in transcriptional gene silencing (Fig. 1b).

A putative DEAH-box RNA helicase (Mut6) has previously been shown to be required for the silencing of transgenes and transposons in *Chlamydomonas* (Wu-Scharf et al. 2000). A mutant strain (Mut6), defective in Mut6, showed enhanced expression and enhanced mobilization of *TOC1*. The transposition activity of *Gulliver* was also higher in the mutant background. Moreover, the stability of *TOC1* transcripts, examined after inhibition of transcription with actinomycin D, was increased in Mut6, clearly implicating the Mut6 machinery in a post-transcriptional mechanism of transposon silencing (Wu-Scharf et al. 2000). The steady-state level of transgenic RNAs and of misspliced *Mut6* transcripts (resulting from the insertion of a mutagenic plasmid into the intron 20 of the *Mut6* gene) was also higher in the Mut6 mutant. Since all these transcripts appear to be aberrant, with defects in proper processing such as splicing and/or polyadenylation, we hypothesized that Mut6 may be part of a surveillance system that controls the expression of genes encoded by transposons and some transgenes because they often produce abnormal RNAs. However, the precise role of the Mut6 putative RNA helicase and its relationship to the RNAi machinery remained unresolved.

We have recently found that Mut6 associates with an RNA binding protein (Rbp1), a staphylococcal nuclease homolog (Snh1), and the glycolytic

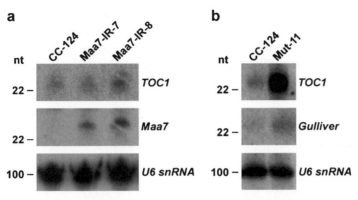

Fig. 1 Transposon and transgene derived small interfering RNAs (*siRNAs*) in *Chlamydomonas*. Column-fractionated small RNAs were separated in a 15% denaturing polyacrylamide gel, electroblotted onto a nylon membrane, and hybridized with the probes indicated, as previously described (Rohr et al. 2004). **a** Small RNAs were isolated from a wild-type *C. reinhardtii* strain (*CC-124*) and from two strains containing inverted repeat transgenes designed to suppress expression of the *Maa7* gene (encoding tryptophan synthase β subunit) (*Maa7-IR-7* and *Maa7-IR-8*) (Rohr et al. 2004). The *panels* show Northern blots of siRNAs sequentially hybridized with probes corresponding to the *TOC1* retrotransposon long terminal repeat or the *Maa7* 3' untranslated region. The same filter was reprobed with the U6 small nuclear RNA (*snRNA*) sequence as a control for equivalent loading of the lanes. Note that the *TOC1* siRNAs are slightly smaller than those corresponding to *Maa7*. **b** Small RNAs were isolated from the wild-type strain (CC-124) and from a mutant strain defective in transcriptional gene silencing (Mut-11) (Zhang et al. 2002). The *panels* show Northern blots of siRNAs sequentially hybridized with probes corresponding to the *Gulliver* DNA transposon right terminus or the *TOC1* retrotransposon long terminal repeat. Transposon-derived siRNAs are present at very low levels in the wild-type strain but they increase substantially in the mutant background. We speculate that an enhancement in overall genome transcription in Mut-11 (due to defects in chromatin repression) results in greater production of both sense transcripts (from the transposon transcription units) as well as antisense RNAs (as read-through transcripts from oppositely oriented neighboring promoters). Sense and antisense RNAs likely anneal producing increased amounts of double-stranded RNA that is processed by Dicer into siRNAs

enzyme enolase (Eno1). These interactions appear to be specific since they occur reproducibly in a yeast two-hybrid system and in Far-Western and co-immunoprecipitation assays (unpublished data). Both Snh1 and Rbp1 bind single-stranded nucleic acids, whereas Snh1 also exhibits Ca^{2+} stimulated nuclease activity, suggesting that they play a role in RNA transactions. The interaction of Mut6 with Snh1 and Rbp1 is also supported by in vivo experiments, since suppression of *Snh1* or *Rbp1* expression by RNAi resulted in reactivation of a post-transcriptionally silenced transgene. Similarly to Mut6, the steady-state RNA level of the *TOC1* retrotransposon as well as its transposition frequency were enhanced in these transgenic lines (unpublished data).

Interestingly, Mut6 as well as the strains where *Snh1* or *Rbp1* expression was suppressed by RNAi also showed reduced levels of *Chlamydomonas* Dicer and Ago1 transcripts (unpublished data). Moreover, consistent with a diminished amount of functional Dicer, *TOC1* siRNAs were also present at lower levels. Comparable phenotypes (i.e., reduced amounts of *TOC1* siRNAs and enhanced levels of *TOC1* transcripts) were observed in *Chlamydomonas* strains where Dicer expression was directly downregulated by RNAi (unpublished data).

The Mut6 protein is most similar to human PRP16, a homolog of an essential precursor mRNA (pre-mRNA) splicing factor in *Saccharomyces cerevisiae*, and to MOG-1, encoded by a gene involved in sex determination in *C. elegans* (Wu-Scharf et al. 2000); however, Mut6 is not necessary for general splicing in *Chlamydomonas* (Wu-Scharf et al. 2000). Rbp1 has an RS domain (rich in repeating Arg-Ser dipeptides) and its general structure resembles that of SR proteins involved in pre-mRNA splicing, mRNA export, and RNA turnover (Graveley 2000). Snh1 has a central domain homologous to the *Staphylococcus aureus* extracellular nuclease, including conservation of the essential amino acids in the catalytic site (Loll and Lattman 1989). Fusions of these proteins with a β-glucuronidase reporter localize either exclusively or partially in the nuclear compartment (unpublished data). Thus, Mut6, Rbp1, Snh1, and, possibly, Eno1 appear to be subunits of a nuclear complex that likely regulates the processing of specific pre-mRNAs, including those corresponding to components of the RNAi machinery such as Dicer and Ago1. Alternatively, in a not mutually exclusive role, the Mut6 complex may also be part of a surveillance system that directly recognizes and degrades improperly processed RNAs such as those of the *TOC1* retroelement and certain transgenes.

In *N. crassa*, the RNAi machinery is not required for heterochromatin formation and DNA methylation (Chicas et al. 2004; Freitag et al. 2004). However, siRNAs derived from transposon relics are common and strains defective in RNAi components show elevated transcript levels corresponding to transposon-related sequences. These observations suggest that the *Neurospora* RNAi machinery plays a role in controlling the expression of transposon and other repetitive sequences post-transcriptionally, by directing the degradation of cognate transcripts (Chicas et al. 2004). In *C. elegans*, silencing of the Tc1 transposon also appears to be (at least in part) post-transcriptional (Sijen and Plasterk 2003; Vastenhouw and Plasterk 2004). Similarly, our findings in the green alga *Chlamydomonas* indicate that the *TOC1* retrotransposon can be suppressed by the targeted degradation of the corresponding RNAs. Therefore, in some eukaryotes, siRNA-directed transcript turnover is indeed one of the mechanisms responsible for the silencing of transposable elements, though, a potential role of miRNAs in the control of transposon activity remains to be explored.

3
Transcriptional Gene Silencing

3.1
General Mechanisms

In eukaryotes, DNA is packaged into the nucleus by assembly into chromatin, the physiological substrate for virtually all DNA-dependent processes. The fundamental unit of chromatin is the nucleosome, consisting of DNA wrapped around an octamer of core histones, containing two molecules each of H2A, H2B, H3, and H4 (Bender 2004). Nucleosomal DNA is further packaged by higher-order folding and association with other proteins (Dorigo et al. 2004; Francis et al. 2004). Chromatin can exist either as heterochromatin, cytologically condensed during interphase, late replicating, and with regular nucleosomal organization, or as euchromatin, decondensed during interphase, early replicating, and characterized by more disorganized nucleosomal arrays (Dillon and Festenstein 2002). A link between heterochromatin and transcriptional silencing has been firmly established by detailed analysis of position-effect variegation in *D. melanogaster* (Wakimoto 1998; Dillon and Festenstein 2002). In contrast, genes in euchromatin are conditionally accessible to the transcription machinery. However, chromatin states are highly dynamic and transcriptionally repressed chromatin can also exist in euchromatic domains (Dillon and Festenstein 2002; Fahrner and Baylin 2003; Rice et al. 2003).

Chromatin states are influenced by postsynthetic modifications of both histones and DNA (Bird 2002; Richards and Elgin 2002; Fischle et al. 2003). Histone residues can be covalently modified by enzymatic complexes that include, among others, acetylation, ADP-ribosylation, methylation, phosphorylation, sumoylation, or ubiquitination activities (Fischle et al. 2003; Sims et al. 2003; Tariq and Paszkowski 2004). In higher eukaryotes, the main postsynthetic DNA change takes the form of cytosine methylation (Bird 2002; Richards and Elgin 2002; Bender 2004). These histone and DNA modifications can affect chromatin states by directly modulating nucleosome structure and/or by creating binding surfaces for chromatin structural/modifying factors (Bird 2002; Richards and Elgin 2002; Fischle et al. 2003; Sims et al. 2003). Moreover, the potential specificity of these signal/recognition interactions led to the proposal of the "histone code" hypothesis, by which certain combinations of histone modifications control the association of proteins with chromatin and determine the functional state of the underlying DNA (Strahl and Allis 2000; Jenuwein and Allis 2001).

Histone modifications can function as epigenetic marks for either active or repressed chromatin (Fischle et al. 2003; Lachner et al. 2003; Sims et al. 2003). In most eukaryotes examined to date, histone H3 lysine 4 (H3K4) dimethylation or trimethylation correlates with transcriptionally permissive chromatin, whereas histone H3 lysine 9 (H3K9) methylation occurs preferentially within silent chromatin (Santos-Rosa et al. 2002; Fischle et al. 2003; Lachner et al. 2003; Sims et al. 2003; Peters et al. 2003; Rice et al. 2003; Tariq and Paszkowski 2004). Histone modifications are also likely to cooperate with DNA methylation in establishing long-term silent chromosomal domains (Bird 2002; Richards and Elgin 2002; Sims et al. 2003; Bender 2004; Tariq and Paszkowski 2004). For instance, pericentromeric heterochromatin is characterized by DNA hypermethylation, histones H3 and H4 hypoacetylation, and by specific methylation of H3K9, predominantly trimethylation in mammals (Peters et al 2003; Rice et al. 2003) and dimethylation in *A. thaliana* (Soppe et al. 2002; Jackson et al. 2004; Tariq and Paszkowski 2004). In several eukaryotes, histone H3K9 methylation and DNA methylation appear to be interdependent, perhaps as a result of interactions between their respective methyltransferase complexes (Lippman and Martienssen 2004). A further level of complexity is introduced by the potential role of RNAs and the RNAi machinery in the establishment of silent chromatin domains (Volpe et al. 2002; Zilberman et al. 2003; Chan et al. 2004; Matzke et al. 2004; Pal-Bhadra et al. 2004).

Components of the RNAi machinery are required for DNA and histone H3K9 methylation in higher plants (Zilberman et al. 2003; Baulcombe 2004; Bender 2004; Chan et al. 2004; Matzke et al. 2004), DNA elimination in *Tetrahymena* (Mochizuki et al. 2002), and heterochromatin formation in (peri)centromeric DNA regions in *Drosophila*, humans, and the fission yeast *Schizosaccharomyces pombe* (Volpe et al. 2002; Fukagawa et al. 2004; Pal-Bhadra et al. 2004). Elegant studies in *S. pombe* have suggested that the stability of centromeric heterochromatin is regulated by specific RNAs that originate as bidirectional transcripts of centromeric repeats and are further processed into siRNAs (Volpe et al. 2002; Sugiyama et al. 2005). Deletion of RNAi effectors, Argonaute (Ago1), Dicer (Dcr1), or an RNA-dependent RNA polymerase (Rdp1), impairs epigenetic silencing at centromeres and the initiation of heterochromatin assembly at the mating type locus in *S. pombe*. An RNA-induced transcriptional silencing (RITS) complex, consisting of Ago1, Chp1 (a chromodomain protein), Tas3 (an uncharacterized factor), and siRNAs, has been shown to be necessary for heterochromatin assembly (Verdel et al. 2004). The RITS complex localizes to all known heterochromatic loci and acts primarily in *cis* to promote transcriptional silencing (Noma et al. 2004). In fact, the RITS effector complex seemingly participates in a self-reinforcing loop mechanism by which siRNAs generated by the RNAi machinery help target heterochromatin proteins such as Clr4 (an H3K9-specific histone methyltransferase) to homologous DNA sequences. The H3K9 methyl mark then directs the binding of chromodomain proteins such as Swi6 (a heterochromatin protein 1 homolog), leading to the formation of heterochromatic structures (Noma et al. 2004). Clr4-mediated H3K9 methylation also anchors the RITS complex to chromatin (likely via the binding of the Chp1 chromodomain), allowing the RNAi machinery to act in *cis* to process efficiently nascent transcripts from repeated sequences (Noma et al. 2004). In *S. pombe*, another complex consisting of Rdp1, Hrr1 (an RNA helicase), and Cid12 (a member of the polyA polymerase family) has recently been shown to interact with the RITS complex and associate with noncoding centromeric RNAs in a Dicer-dependent manner (Motamedi et al. 2004). This complex is essential for the efficient generation of siRNAs (Motamedi et al. 2004; Sugiyama et al. 2005). Thus, the data for *S. pombe* suggest a self-reinforcing cycle by which heterochromatic marks originally established by the RNAi pathway maintain the RITS complex and the Rdp1 complex at specific loci, stabilizing heterochromatin formation.

Despite the clear role of RNA-mediated processes in heterochromation formation in some organisms, there is also evidence for the establishment of repressive chromatin domains independently of RNAi. For instance, the RNAi machinery is not required for heterochromatin formation and DNA methylation in *Neurospora* (Chicas et al. 2004; Freitag et al. 2004). Similarly, in *S. pombe*, activating transcription factor/cyclic AMP response-element binding protein (ATF/CREB) family proteins can nucleate heterochromatin assembly at the mating type region independently of RNAi (Jia et al. 2004). Moreover, repressive chromatin also occurs in the budding yeast *Saccharomyces cerevisiae* (Perrod and Gasser 2003; Zhang and Reese 2004), which lacks essential components of the RNAi machinery (Sigova et al. 2004). Thus, transcriptional gene silencing in eukaryotes is achieved via the formation and the propagation of specific chromatin states, characterized by certain histone modifications and associated (hetero)chromatic proteins. In some species, such as mammals and higher plants, heterochromatin is also characterized by the hypermethylation of DNA at cytosine residues; however, silent chromatin states appear to be established and/or maintained in different ways, in some cases, but not always, depending on the RNAi machinery.

3.2
TGS and Transposon Silencing

In higher plants, TGS was first recognized owing to the silencing of introduced transgenes, but it also controls the activity of endogenous genes and transposons (Matzke et al. 2001; Okamoto and Hirochika 2001; Vaucheret and Fagard 2001; Bender 2004; Lippman and Martienssen 2004). In the case of transgenes, TGS often involves cytosine methylation of promoter regions and reduced accessibility to DNase I, suggesting an altered chromatin structure (Vaucheret and Fagard 2001; Bender 2004). Moreover, dsRNA and/or siRNAs derived from promoter regions have been implicated in the DNA methylation and transcriptional inactivation of homologous sequences (Mette et al. 2000; Matzke et al. 2001). This RNA-directed DNA methylation (RdDM) is highly sequence specific, since methylation does not spread significantly beyond the boundary of the RNA trigger (Pélissier et al. 1999; Bender 2004; Matzke et al. 2004). In *Chlamydomonas* and other volvocine algae, silenced multiple-copy transgenes also exhibit high levels of DNA methylation (Cerutti et al. 1997; Babinger et al. 2001). In contrast, single-copy transgenes are subject to TGS without detectable cytosine methylation (Cerutti et al. 1997).

The limited data on the activity of plant transposons suggest that transposition is controlled primarily at the level of transcription (Feschotte et al. 2002). Indeed, many transposable elements appear to be regulated by the same mechanisms as those required for RdDM of transgenes (Lippman et al. 2003; Bender 2004; Lippman and Martienssen 2004). In *Arabidopsis*, analysis of a number of mutants revealed that both CG and non-CG DNA methylation, as well as histone deacetylation, and histone H3K9 methylation contribute to transposon silencing, although to different extents depending on the specific transposable element (Gendrel et al. 2002; Kato et al. 2003; Lippman et al. 2003; Lippman and Martienssen 2004; Lippman et al. 2004). RNAi likely plays a role in the sequence-specific targeting of these transposons and other repeat sequences for heterochromatin formation. Many endogenous siRNAs in *Arabidopsis* correspond to silenced transposons, and, as for *S. pombe* heterochromatic repeats, an RdRP (RDR2) and a Dicer homolog (DCL3) are required for the production of transposon-derived siRNAs (Llave et al. 2002; Lippman et al. 2004; Xie et al. 2004). However, only a handful of *Arabidopsis* transposons are activated in mutants of the RNAi machinery (Zilberman et al. 2003; Lippman et al. 2003; Lippman and Martienssen 2004), even when siRNAs are completely lost (Xie et al. 2004). Although in some cases this lack of an effect may be due to genetic redundancy, the evidence suggests that RNAi may be involved in the establishment of transposon silencing in *Arabidopsis*, as demonstrated for *FWA* transgenes (encoding a homeodomain-containing transcription factor that regulates flowering) (Chan et al. 2004), whereas maintenance of transposon repression can occur by other means (Lippman and Martienssen 2004). However, the possibility that RNAi has a redundant role in silencing maintenance (Bender 2004) and/or the existence of transcriptional silencing mechanisms entirely independent of the RNAi machinery cannot be ruled out.

In *Chlamydomonas*, two genes have been implicated in the transcriptional silencing of transposons and single-copy transgenes: *Mut9*, which encodes a novel serine/threonine protein kinase with a catalytic domain similar to that of casein kinase I (Jeong et al. 2002), and *Mut11*, which encodes a WD40-repeat containing protein (Zhang et al. 2002). Mutant strains deleted for these genes are defective in the silencing of the retrotransposon *TOC1* and the DNA transposon *Gulliver* (Jeong et al. 2002; Zhang et al. 2002). Interestingly, in a double mutant (*mut9 mut11*), *Gulliver* seems to transpose at a much higher frequency than in either of the single mutants (Jeong et al. 2002). In contrast to most transposons examined in *Arabidopsis*, which are predominantly located in heterochromatic regions (Lippman and Martienssen

2004; Lippman et al. 2004), *TOC1* and *Gulliver* are present in limited numbers (10–40 copies per haploid genome), dispersed throughout the nuclear genome, and do not appear to be methylated (Day et al. 1988; Ferris 1989; Hall and Luck 1995; unpublished data). In fact these transposable elements appear to integrate in euchromatic, intergenic regions (http://genome.jgi-psf.org/chlre2/chlre2.home.html) and were first isolated owing to the mutant phenotypes caused by their insertion into active genes (Day et al. 1988; Schnell and Lefebvre 1993).

Recombinant Mut9 phosphorylates histones H3 and H2A in vitro and genetic analyses strongly suggest that this kinase is involved in placing an epigenetic mark that targets specific chromosomal domains for transcriptional repression (unpublished data). However, the actual molecular role(s) of Mut9 is currently unknown. Mut11 is related to a conserved subunit of trithorax-like histone methyltransferase complexes. In yeast two-hybrid assays, Mut11 interacts with a SET domain protein, Set1, capable of methylating lysine 4 of histone H3 (van Dijk et al. 2005). A Mut11 complex, isolated by affinity chromatography, contains homologs of human Rbbp5 (retinoblastoma binding protein 5) and Ash2L (absent, small or homeotic discs 2-like) (van Dijk et al. 2005), conserved subunits of histone methyltransferase complexes (Hughes et al. 2004). Moreover, the purified Mut11 complex methylates histones H3, H2A, and H4 in vitro. *Set1* RNAi-suppressed strains and a Mut11 insertional mutant show defects in transcriptional silencing of transgenes and transposons, as well as a global loss of monomethyl H3K4 (van Dijk et al. 2005). By chromatin immunoprecipitation analyses, these strains also displayed substantial reduction in monomethylated H3K4 associated with transcriptionally derepressed transgenes and the *TOC1* retrotransposon. In contrast, the mutant and RNAi strains showed no defects in the H3K4 methylation pattern associated with a transcriptionally active ribosomal protein gene (van Dijk et al. 2005). These observations suggest that the Mut11 machinery is necessary for the maintenance of repressed euchromatic domains in *Chlamydomonas*, via the monomethylation of H3K4 and perhaps other modifications in histones H2A and H4.

The transcriptional silencing mediated by the Mut11 machinery does not appear to be dependent on RNAi. RNAi-suppression of Dicer in a wild-type *Chlamydomonas* strain did not affect the H3K4 methylation pattern of repressed transgenes and the *TOC1* retrotransposon. Moreover, downregulation of Dicer by RNAi in the Mut11 mutant background resulted in a synergistic increase in *TOC1* expression and transposition (unpublished data). This lack of epistatic effects indicates that Mut11 and Dicer function in at least partly independent pathways of transposon silencing. Thus, in *Chlamydomonas*, the transcriptional repression of transposons can occur, as in other eukaryotes, by pathways involving histone modifications and changes in chromatin structure. However, the characterized TGS mechanisms also reveal apparent differences. For instance, H3K4 monomethylation is clearly involved in transposon silencing in *Chlamydomonas*, whereas the role of this histone modification in other organisms remains unexplored. Conversely, the RNAi machinery has been demonstrated to participate in heterochromatin formation and transposon repression in a number of eukaryotes, but its function in *Chlamydomonas* TGS, if any, is currently unknown. Likewise, although DNA methylation plays a role in the silencing of multiple-copy transgenes in volvocine algae, its potential function in the suppression of transposon activity awaits further examination.

4
Concluding Remarks

Transposons are infectious agents in sexual populations and can go to fixation even if they reduce the fitness of the host. Sexual hosts are therefore under selective pressure to evolve defensive functions that alleviate the fitness penalty imposed by active transposable elements (Bestor 2003). Many eukaryotes have evolved epigenetic silencing mechanisms that play a role in controlling transposon activity. The unicellular green alga *C. reinhardtii* appears to have several, at least partly independent, transposon repression pathways that operate at either the transcriptional or the post-transcriptional level. Transcriptional silencing requires specific histone modifications (such as H3K4 monomethylation) for the formation of repressive chromatin states, whereas post-transcriptional silencing involves the RNAi machinery and targeted RNA degradation. Mechanistic connections between these pathways, although clearly occurring in several organisms, have not been detected in *Chlamydomonas*. Thus, the limited data on transposon suppression suggest that multiple silencing mechanisms may operate as a defense system against the massive expansion of these elements in eukaryotes. Moreover, independent silencing pathways likely increase the genetic robustness of transposon repression against mutational and environmental perturbations.

References

Ambros V (2004) The functions of animal microRNAs. Nature 431:350–355
Ambros V, Lee RC, Lavanway A, Williams PT, Jewell D (2003) MicroRNAs and other tiny endogenous RNAs in *C. elegans*. Curr Biol 13:807–818
Aravin AA, Naumova NM, Tulin AV, Vagin VV, Rozovsky YM, Gvozdev VA (2001) Double-stranded RNA-mediated silencing of genomic tandem repeats and transposable elements in the *D. melanogaster* germline. Curr Biol 11:1017–1025
Aravin AA, Lagos-Quintana M, Yalcin A, Zavolan M, Marks D, Snyder B, Gaasterland T, Meyer J, Tuschl T (2003) The small RNA profile during Drosophila melanogaster development. Dev Cell 5:337–350

Babinger P, Kobl I, Mages W, Schmitt R (2001) A link between DNA methylation and epigenetic silencing in transgenic *Volvox carteri*. Nucleic Acids Res 29:1261–1271

Bartel DP (2004) MicroRNAs: genomics, biogenesis, mechanism, and function. Cell 116:282–297

Baulcombe D (2004) RNA-silencing in plants. Nature 431:356–363

Bender J (2004) Chromatin-based silencing mechanisms. Curr Opin Plant Biol 7:521–526

Bestor TH (2003) Cytosine methylation mediates sexual conflict. Trends Genet 19:185–190

Bird A (2002) DNA methylation patterns and epigenetic memory. Genes Dev 16:6–21

Carmell MA, Xuan Z, Zhang MQ, Hannon GJ (2002) The Argonaute family: tentacles that reach into RNAi, developmental control, stem cell maintenance, and tumorigenesis. Genes Dev 16:2733–2742

Carrington JC, Ambros V (2003) Role of microRNAs in plant and animal development. 301:336–338

Cerutti H (2003) RNA interference: traveling in the cell and gaining functions? Trends Genet 19:39–46

Cerutti H, Johnson AM, Gillham NW, Boynton JE (1997) Epigenetic silencing of a foreign gene in nuclear transformants of *Chlamydomonas*. Plant Cell 9:925–945

Chan SWL, Zilberman D, Xie Z, Johansen LK, Carrington JC, Jacobsen SE (2004) RNA silencing genes control de novo DNA methylation. Science 303:1336

Chicas A, Cogoni C, Macino G (2004) RNAi-dependent and RNAi-independent mechanisms contribute to the silencing of RIPed sequences in *Neurospora crassa*. Nucleic Acids Res 32:4237–4243

Cogoni C, Macino G (2000) Post-transcriptional gene silencing across kingdoms. Curr Opin Genet Dev 10:638–643

Colombo R, Bignamini AA, Carobene A, Sasaki J, Tachikawa M, Kobayashi K, Toda T (2000) Age and origin of the FCMD 3′-untranslated-region retrotransposal insertion mutation causing Fukuyama-type congenital muscular dystrophy in the Japanese population. Hum Genet 107:559–567

Day A, Rochaix JD (1991) Structure and inheritance of sense and anti-sense transcripts from a transposon in the green alga *Chlamydomonas reinhardtii*. J Mol Biol 218:273–291

Day A, Schirmer-Rahire M, Kuchka MR, Mayfield SP, Rochaix JD (1988) A transposon with an unusual arrangement of long terminal repeats in the green alga *Chlamydomonas reinhardtii*. EMBO J 7:1917–1927

Denli AM, Hannon GJ (2003) RNAi: an ever-growing puzzle. Trends Biochem Sci 28:196–201

Dillon N, Festenstein R (2002) Unravelling heterochromatin: competition between positive and negative factors regulates accessibility. Trends Genet 18:252–258

Djikeng A, Shi H, Tschudi C, Ullu E (2001) RNA interference in *Trypanosoma brucei*: cloning of small interfering RNAs provides evidence for retroposon-derived 24–26 nucleotide RNAs. RNA 7:1522–1530

Doench JG, Petersen CP, Sharp PA (2003) siRNAs can function as miRNAs. Genes Dev 17:438–442

Dorigo B, Schalch T, Kulangara A, Duda S, Schroeder RR, Richmond TJ (2004) Nucleosome arrays reveal the two-start organization of the chromatin fiber. Science 306:1571–1573.

Fahrner JA, Baylin SB (2003) Heterochromatin: stable and unstable invasions at home and abroad. Genes Dev 17:1805–1812

Ferris PJ (1989) Characterization of a *Chlamydomonas* transposon, *Gulliver*, resembling those in higher plants. Genetics 122:363–377

Feschotte C, Jiang N, Wessler SR (2002) Plant transposable elements: where genetics meets genomics. Nat Rev Genet 3:329–341

Fire A (1999) RNA-triggered gene silencing. Trends Genet 15:358–363

Fischle W, Wang YM, Allis CD (2003) Histone and chromatin cross-talk. Curr Opin Cell Biol 15:172–183

Francis NJ, Kinston RE, Woodcock CL (2004) Chromatin compaction by a polycomb group protein complex. Science 306:1574–1577

Freitag M, Lee DW, Kothe GO, Pratt RJ, Aramayo R, Selker EU (2004) DNA methylation is independent of RNA interference in *Neurospora*. Science 304:1939

Fukagawa T, Nogami M, Yoshikawa M, Ikeno M, Okazaki T, Takami Y, Nakayama T, Oshimura M (2004) Dicer is essential for formation of the heterochromatin structure in vertebrate cells. Nat Cell Biol 6:784–791

Gendrel AV, Lippman Z, Yordan C, Colot V, Martienssen RA (2002) Dependence of heterochromatic histone H3 methylation patterns on the *Arabidopsis* gene *DDM1*. Science 297:1871–1873

Graveley BR (2000) Sorting out the complexity of SR protein functions. RNA 6:1197–1211

Haley B, Zamore PD (2004) Kinetic analysis of the RNAi enzyme complex. Nat Struct Mol Biol 11:599–606

Hall JL, Luck DJL (1995) Basal body-associated DNA: *In situ* studies in *Chlamydomonas reinhardtii*. Proc Natl Acad Sci USA 92:5129–5133

Havecker ER, Gao X, Voytas DF (2004) The diversity of LTR retrotransposons. Genome Biol 5:225

Hughes CM, Rozenblatt-Rosen O, Milne TA, Copeland TD, Levine SS, Lee JC, Hayes DN, Shanmugam KS, Bhattacharjee A, Biondi CA, Kay GF, Hayward NK, Hess JL, Meyerson M (2004) Menin associates with a trithorax family histone methyltransferase complex and with the Hoxc8 locus. Mol Cell 13:587–597

Hutvágner G, Zamore PD (2002) A microRNA in a multiple-turnover RNAi enzyme complex. Science 297:2056–2060

Jackson JP, Johnson L, Jasencakova Z, Zhang X, PerezBurgos L, Singh PB, Cheng XD, Schubert I, Jenuwein T, Jacobsen SE (2004) Dimethylation of histone H3 lysine 9 is a critical mark for DNA methylation and gene silencing in Arabidopsis thaliana. Chromosoma 112:308–315

Jenuwein T, Allis CD (2001) Translating the histone code. Science 293:1074–1080

Jeong B-r, Wu-Scharf D, Zhang C, Cerutti H (2002) Suppressors of transcriptional transgenic silencing in *Chlamydomonas* are sensitive to DNA-damaging agents and can reactivate transposable elements. Proc Natl Acad Sci USA 99:1076–1081

Jia S, Noma K-i, Grewal SIS (2004) RNAi-independent heterochromatin nucleation by the stress-activated ATF/CREF family proteins. Science 304:1971–1976

Jorgensen RA, Que QD, Stam M (1999) Do unintended antisense transcripts contribute to sense cosuppression in plants? Trends Genet 15:11–12

Kato M, Miura A, Bender J, Jacobsen SE, Kakutani T (2003) Role of CG and non-CG methylation in immobilization of transposons in *Arabidopsis*. Curr Genet 13:421–426

Lachner M, O'Sullivan RJ, Jenuwein T (2003) An epigenetic road map for histone lysine methylation. J Cell Sci 116:2117–2124

Lander ES, Linton LM, Birren B, Nusbaum C, Zody MC, Baldwin J, Devon K, Dewar K, Doyle M, FitzHugh W, Funke R, Gage D, Harris K, Heaford A, Howland J, Kann L, Lehoczky J, LeVine R, McEwan P, McKernan K, Meldrim J, Mesirov JP, Miranda C, Morris W, Naylor J, Raymond C, Rosetti M, Santos R, Sheridan A, Sougnez C, Stange-Thomann N, Stojanovic N, Subramanian A, Wyman D, Rogers J, Sulston J, Ainscough R, Beck S, Bentley D, Burton J, Clee C, Carter N, Coulson A, Deadman R,

Deloukas P, Dunham A, Dunham I, Durbin R, French L, Grafham D, Gregory S, Hubbard T, Humphray S, Hunt A, Jones M, Lloyd C, McMurray A, Matthews L, Mercer S, Milne S, Mullikin JC, Mungall A, Plumb R, Ross M, Shownkeen R, Sims S, Waterston RH, Wilson RK, Hillier LW, McPherson JD, Marra MA, Mardis ER, Fulton LA, Chinwalla AT, Pepin KH, Gish WR, Chissoe SL, Wendl MC, Delehaunty KD, Miner TL, Delehaunty A, Kramer JB, Cook LL, Fulton RS, Johnson DL, Minx PJ, Clifton SW, Hawkins T, Branscomb E, Predki P, Richardson P, Wenning S, Slezak T, Doggett N, Cheng JF, Olsen A, Lucas S, Elkin C, Uberbacher E, Frazier M, Gibbs RA, Muzny DM, Scherer SE, Bouck JB, Sodergren EJ, Worley KC, Rives CM, Gorrell JH, Metzker ML, Naylor SL, Kucherlapati RS, Nelson DL, Weinstock GM, Sakaki Y, Fujiyama A, Hattori M, Yada T, Toyoda A, Itoh T, Kawagoe C, Watanabe H, Totoki Y, Taylor T, Weissenbach J, Heilig R, Saurin W, Artiguenave F, Brottier P, Bruls T, Pelletier E, Robert C, Wincker P, Smith DR, Doucette-Stamm L, Rubenfield M, Weinstock K, Lee HM, Dubois J, Rosenthal A, Platzer M, Nyakatura G, Taudien S, Rump A, Yang H, Yu J, Wang J, Huang G, Gu J, Hood L, Rowen L, Madan A, Qin S, Davis RW, Federspiel NA, Abola AP, Proctor MJ, Myers RM, Schmutz J, Dickson M, Grimwood J, Cox DR, Olson MV, Kaul R, Raymond C, Shimizu N, Kawasaki K, Minoshima S, Evans GA, Athanasiou M, Schultz R, Roe BA, Chen F, Pan H, Ramser J, Lehrach H, Reinhardt R, McCombie WR, de la Bastide M, Dedhia N, Blocker H, Hornischer K, Nordsiek G, Agarwala R, Aravind L, Bailey JA, Bateman A, Batzoglou S, Birney E, Bork P, Brown DG, Burge CB, Cerutti L, Chen HC, Church D, Clamp M, Copley RR, Doerks T, Eddy SR, Eichler EE, Furey TS, Galagan J, Gilbert JG, Harmon C, Hayashizaki Y, Haussler D, Hermjakob H, Hokamp K, Jang W, Johnson LS, Jones TA, Kasif S, Kaspryzk A, Kennedy S, Kent WJ, Kitts P, Koonin EV, Korf I, Kulp D, Lancet D, Lowe TM, McLysaght A, Mikkelsen T, Moran JV, Mulder N, Pollara VJ, Ponting CP, Schuler G, Schultz J, Slater G, Smit AF, Stupka E, Szustakowski J, Thierry-Mieg D, Thierry-Mieg J, Wagner L, Wallis J, Wheeler R, Williams A, Wolf YI, Wolfe KH, Yang SP, Yeh RF, Collins F, Guyer MS, Peterson J, Felsenfeld A, Wetterstrand KA, Patrinos A, Morgan MJ, de Jong P, Catanese JJ, Osoegawa K, Shizuya H, Choi S, Chen YJ; International Human Genome Sequencing Consortium (2001) Initial sequencing and analysis of the human genome. Nature 409:860–921

Lippman Z, Martienssen R (2004) The role of RNA interference in heterochromatic silencing. Nature 431:364–370

Lippman Z, May B, Yordan C, Singer T, Martienssen R (2003) Distinct mechanisms determine transposon inheritance and methylation via small interfering RNA and histone modification. PLoS Biol 1:E67

Lippman Z, Gendrel A-V, Black M, Vaughn MW, Dedhia N, McCombie WR, Lavine K, Mittal V, May B, Kasschau KD, Carrington JC, Doerge RW, Colot V, Martienssen R (2004) Role of transposable elements in heterochromatin and epigenetic control. Nature 430:471–476

Liu J, Carmell MA, Rivas FV, Marsden CG, Thomson JM, Song JJ, Hammond SM, Joshua-Tor L, Hannon GJ (2004) Argonaute2 is the catalytic engine of mammalian RNAi. Science 305:1437–1441

Llave C, Kasschau KD, Rector MA, Carrington JC (2002) Endogenous and silencing-associated small RNAs in plants. Plant Cell 14:1605–1619

Loll PJ, Lattman EE (1989) The crystal structure of the ternary complex of Staphylococcal nuclease, Ca2+, and the inhibitor pdTp, refined at 1.65 Å. Proteins 5:183–201

Lower R (1999) The pathogenic potential of endogenous retroviruses: facts and fantasies. Trends Microbiol 7:350–356

Martens H, Novotny J, Oberstrass J, Steck TL, Postlethwait P, Nellen W (2002) RNAi in *Dictyostelium*: the role of RNA-directed RNA polymerases and double-stranded RNase. Mol Biol Cell 13:445–453

Matzke M, Matzke AJM, Kooter JM (2001) RNA: guiding gene silencing. Science 293:1080–1083

Matzke M, Aufsatz W, Kanno T, Daxinger L, Papp I, Mette MF, Matzke AJ (2004) Genetic analysis of RNA-mediated transcriptional gene silencing. Biochim Biophys Acta 15:129–141

McClintock B (1984) The significance of responses of the genome to challenge. Science 226:792–801

Meister G, Tuschl T (2004) Mechanisms of gene silencing by double-stranded RNA. Nature 431:343–349

Mette MF, Aufsatz W, van der Winden J, Matzke MA, Matzke AJ (2000) Transcriptional silencing and promoter methylation triggered by double-stranded RNA. EMBO J 19:5194–5201

Miki Y (1998) Retrotransposal integration of mobile genetic elements in human diseases. J Hum Genet 43:77–84

Mochizuki K, Fine NA, Fujisawa T, Gorovsky MA (2002) Analysis of a piwi-related gene implicates small RNAs in genome rearrangement in tetrahymena. Cell 110:689–699

Motamedi MR, Verdel A, Colmenares SU, Gerber SA, Gygi SP, Moazed D (2004) Two RNAi complexes, RITS and RDRC, physically interact and localize to noncoding centromeric RNA. Cell 119:789–802

Nekrutenko A, Li W-H (2001) Transposable elements are found in a large number of human protein-coding genes. Trends Genet 17:619–621

Noma K, Sugiyama T, Cam H, Verdel A, Zofall M, Jia S, Moazed D, Grewal SIS (2004) RITS acts in *cis* to promote RNA interference-mediated transcriptional and post-transcriptional silencing. Nat Genet 36:1174–1180

Okamoto H, Hirochika H (2001) Silencing of transposable elements in plants. Trends Plant Sci 6:527–534

Pal-Bhadra M, Leibovitch BA, Gandhi SG, Rao M, Bhadra U, Birchler JA, Elgin SCR (2004) Heterochromatic silencing and HP1 localization in Drosophila are dependent on the RNAi machinery. Science 303:669–672

Parker JS, Roe SM, Barford D (2004) Crystal structure of a PIWI protein suggests mechanisms for siRNA recognition and slicer activity. EMBO J 23:4727–4737

Pélissier T, Thalmeir S, Kempe D, Sänger HL, Wasseneger M (1999) Heavy *de novo* methylation at symmetrical and non-symmetrical sites is a hallmark of RNA-directed DNA methylation. Nucleic Acids Res 27:1625–1634

Perrod S, Gasser SM (2003) Long-range silencing and position effects at telomeres and centromeres: parallels and differences. Cell Mol Life Sci 60:2303–2318

Peters AHFM, Kubicek S, Mechtler K, O'Sullivan RJ, Derijck AAHA, Perez-Burgos L, Kohlmaier A, Opravil S, Tachibana M, Shinkai Y, Martens JHA, Jenuwein T (2003) Partitioning and plasticity of repressive histone methylation states in mammalian chromatin. Mol Cell 12:1577–1589

Pham JW, Pellino JL, Lee YS, Carthew RW, Sontheimer EJ (2004) A dicer-2-dependent 80S complex cleaves targeted mRNAs during RNAi in *Drosophila*. Cell 117:83–93

Rice JC, Briggs SD, Ueberheide B, Barber CM, Shabanowitz J, Hunt DF, Shinkai Y, Allis CD (2003) Histone methyltransferases direct different degrees of methylation to define distinct chromatin domains. Mol Cell 12:1591–1598

Richards EJ, Elgin SCR (2002) Epigenetic codes for heterochromatin formation and silencing: Rounding up the usual suspects. Cell 108:489–500

Rohr J, Sarkar N, Balenger S, Jeong B-r, Cerutti H (2004) Tandem inverted repeat system for selection of effective transgenic RNAi strains in *Chlamydomonas*. Plant J 40:611–621

Rudenko GN, Ono A, Walbot V (2003) Initiation of silencing of maize MuDR/Mu transposable elements. Plant J 33:1013–1025

SanMiguel P, Bennetzen JL (1998) Evidence that a recent increase in maize genome size was caused by the massive amplification of intergene retrotransposons. Ann Bot 82:37–44

Santos-Rosa H, Schneider R, Bannister AJ, Sherriff J, Bernstein BE, Emre NCT, Schreiber SL, Mellor J, Kouzarides T (2002). Active genes are tri-methylated at K4 of histone H3. Nature 419:407–411

Schnell RA, Lefebvre PA (1993) Isolation of the *Chlamydomonas* regulatory gene *NIT2* by transposon tagging. Genetics 134:737–747

Sigova A, Rhind N, Zamore PD (2004) A single Argonaute protein mediates both transcriptional and posttranscriptional silencing in *Schizosaccharomyces pombe*. Genes Dev 18:2359–2367

Sijen T, Plasterk RH (2003) Transposon silencing in the *Caenorhabditis elegans* germ line by natural RNAi. Nature 426:310–314

Sijen T, Fleenor J, Simmer F, Thijssen KL, Parrish S, Timmons L, Plasterk RH, Fire A (2001) On the role of RNA amplification in dsRNA-triggered gene silencing. Cell 107:465–476

Sims RJ, Nishioka K, Reinberg D (2003) Histone lysine methylation: a signature for chromatin function. Trends in Genet 19:629–639

Song JJ, Smith SK, Hannon GJ, Joshua-Tor L (2004) Crystal structure of Argonaute and its implications for RISC slicer activity. Science 305:1434–1437

Soppe WJJ, Jasencakova Z, Houben A, Kakutani T, Meister A, Huang HS, Jacobsen SE, Schubert I, Fransz PF (2002) DNA methylation controls histone H3 lysine 9 methylation and heterochromatin assembly in Arabidopsis. EMBO J 21:6549–6559

Strahl BD, Allis CD (2000) The language of covalent histone modifications. Nature 403:41–45

Sugiyama T, Cam H, Verdel A, Moazed D, Grewal SIV (2005) RNA-dependent RNA polymerase is an essential component of a self-enforcing loop coupling heterochromatin assembly to siRNA production. Proc Natl Acad Sci USA 102:152–157

Tariq M, Paszkowski J (2004) DNA and histone methylation in plants. Trends Genet 20:244–251

Tomari Y, Du T, Haley B, Schwarz DS, Bennett R, Cook HA, Koppetsch BS, Theurkauf WE, Zamore PD (2004) RISC assembly defects in the *Drosophila* RNAi mutant armitage. Cell 116:831–841

van Dijk K, Marley KE, Jeong B-r, Xu J, Hesson J, Cerny RL, Waterborg JH, Cerutti H (2005) Monomethyl histone H3 lysine 4 as an epigenetic mark for silenced chromatin in *Chlamydomonas*. Plant Cell 17:2439–2453

Vastenhouw NL, Plasterk RHA (2004) RNAi protects the *Caenorhabditis elegans* germline against transposition. Trends Genet 20:314–319

Vaucheret H, Fagard M (2001) Transcriptional gene silencing in plants: targets, inducers and regulators. Trends Genet 17:29–35

Verdel A, Jia S, Gerber S, Sugiyama T, Gygi S, Grewal SIS, Moazed D (2004) RNAi-mediated targeting of heterochromatin by the RITS complex. Science 303:672–676

Volpe TA, Kidner C, Hall IM, Teng G, Grewal SIS, Martienssen RA (2002) Regulation of heterochromatic silencing and histone H3 lysine-9 methylation by RNAi. Science 297:1833–1837

Wakimoto BT (1998) Beyond the nucleosome: Epigenetic aspects of position-effect variegation in Drosophila. Cell 93:321–324

Wang S-C, Schnell RA, Lefebvre PA (1998) Isolation and characterization of a new transposable element in *Chlamydomonas reinhardtii*. Plant Mol Biol 38:681–687

Wu-Scharf D, Jeong B-r, Zhang C, Cerutti H (2000) Transgene and transposon silencing in
Chlamydomonas reinhardtii by a DEAH-box RNA helicase. Science 290:1159–1162

Xie Z, Johansen LK, Gustafson AM, Kasschau KD, Lellis AD, Zilberman D, Jacobsen SE,
Carrington JC (2004) Genetic and functional diversification of small RNA pathways in
plants. PLoS Biol 2:E104

Zamore PD (2002) Ancient pathways programmed by small RNAs. Science 296:1265–1269

Zhang C, Wu-Scharf D, Jeong B-r, Cerutti H (2002) A WD40-repeat containing protein,
similar to a fungal co-repressor, is required for transcriptional gene silencing in
Chlamydomonas. Plant J 31:25–36

Zhang Z, Reese JC (2004) Redundant mechanisms are used by Ssn6-Tup1 in repressing
chromosomal gene transcription in *Saccharomyces cerevisiae*. J Biol Chem 279:39240–
39250

Zilberman D, Cao XF, Jacobsen SE (2003) ARGONAUTE4 control of locus-specific siRNA
accumulation and DNA and histone methylation. Science 299:716–719

Nucleic Acids and Molecular Biology, Vol. 17
Wolfgang Nellen, Christian Hammann (Eds.)
Small RNAs
© Springer-Verlag Berlin Heidelberg 2005

RNA-Dependent Gene Silencing and Epigenetics in Animals

Martina Paulsen (✉) · Sascha Tierling · Stephanie Barth · Jörn Walter

FR 8.3 Biowissenschaften—Genetik/Epigenetik, Universität des Saarlandes,
Postfach 151150, 66041 Saarbrücken, Germany
m.paulsen@mx.uni-saarland.de

Abstract In animals noncoding RNAs are involved in a large variety of gene silencing mechanisms. These include post-transcriptional RNA interference (RNAi) that is mediated by small double-stranded RNAs and results in degradation of messenger RNAs as well as epigenetic silencing of genes. RNAi as a naturally occurring silencing mechanism has been well investigated in various eukaryotic organisms. Sequencing of the human and mouse genomes and careful analyses of the related transcriptomes led to the identification of some hundred microRNAs that might regulate endogenous gene expression by RNAi-like mechanisms or repression of translation. In mammals the major protein components of the RNAi machinery have been identified, and RNAi has become a tool for artificial gene silencing in mammalian systems. There is also evidence that in mammalian cells genes can be regulated by noncoding antisense transcripts that are transcribed from the opposite DNA strand. Besides their potential roles in RNAi and repression of translation, short double-stranded RNAs and also long noncoding RNAs are involved in epigenetic gene silencing. In this chapter we give an overview of prominent features of naturally occurring RNAi and also of the potential role of RNAs in epigenetic gene silencing mechanisms in animals, especially in mammals.

1
MicroRNAs and Small Interfering RNAs—
Two Different Types of Small RNAs

Two major pathways of RNA-mediated posttranscriptional silencing have been described for animal systems. Firstly, small 21–25-nt-long antisense RNAs can bind to messenger RNAs (RNAs) and can introduce mRNA degradation (Elbashir et al. 2001). This phenomenon is commonly known as RNA interference (RNAi). The RNAi response is evolutionarily conserved and it is assumed to reflect primarily an endogenous defense mechanism against virus infection or parasitic nucleic acids. This is indicated by the finding that deletion of RNAi components indeed compromises virus resistance in plants (Baulcombe 1999). Secondly, binding of small RNAs to protein-encoding mRNAs can inhibit translation. The choice which of the two processes, i.e. degradation or inhibition of translation, is induced depends on the similarity of the small antisense RNA to its target mRNA (Hutvagner and Zamore 2002; Yekta et al. 2004). In case of a perfect match the target mRNA will be degraded; if the RNA duplex formed contains some mismatches translation will be inhibited.

In general, two different major groups of small RNAs that are involved in posttranscriptional silencing can be distinguished: microRNAs (miRNAs) that are typically about 22 nt in length and small interfering RNAs (siRNAs) that are approximately 21–25-nt long. Both types of short RNAs are generated by the endonucleolytic cleavage of double-stranded RNA (dsRNA) structures into miRNA or siRNA duplexes. However, miRNAs and siRNAs are derived from different substrates that are cut by the RNAse III like enzyme Dicer. Whereas siRNAs are generated from long dsRNAs, miRNAs are products of endogenous, noncoding genes whose transcripts form small stem-loops from which mature miRNAs are cleaved (Fig. 1) (Hutvagner et al. 2001). Though in the mouse and human genomes more than 200 miRNA encoding sequences have been identified (Lagos-Quintana et al. 2001, 2002; Dostie et al. 2003), only for a few of them have potential target genes been suggested. In general, miRNAs appear to be encoded by genes that are not the target genes; hence, miRNAs are usually *trans*-acting factors involved in gene regulation. Exceptions where the miRNA might be encoded by the antisense strand of its potential target transcript are the murine miRNAs *miR*-127 and *miR*-136 that reside on the antisense strand of the retrotransposon-like *Rtl1* gene on chromosome 12 in mice (Seitz et al. 2003). In addition, miRNAs can also be

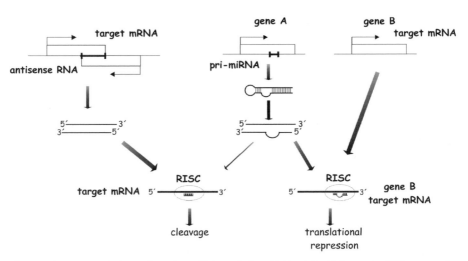

Fig. 1 MicroRNAs (*miRNAs*) and small interfering RNAs (*siRNAs*) act in different pathways of posttranscriptional gene regulation. siRNA precursors are generated by hybridization of two complementary RNAs that are derived from sense and antisense transcription of the target gene. In contrast, miRNA precursors (*pri-miRNAs*) are transcribed from a gene A that is different from the target gene B. After processing siRNAs interact with perfectly matching target messenger RNAs (*mRNAs*) and induce mRNA cleavage. miRNAs pair with the 3′ untranslated region of the target mRNAs of gene B. The not perfectly matching RNA duplex inhibits the translation of gene B. *RISC* RNA-induced silencing complex

produced by viruses, as has been shown for the Epstein–Barr virus in human cells (Pfeffer et al. 2004). Many of the miRNAs are encoded by singular DNA copies that are evenly distributed throughout the genome; however, some are organized as clusters of neighboring miRNA encoding tandem repeats (Seitz et al. 2004; Lagos-Quintana et al. 2001).

Experimentally introduced long dsRNAs are able to induce RNAi, indicating that artificial dsRNAs are also processed into siRNAs by the intracellular RNAi machinery (Fire et al. 1998; Elbashir et al. 2001; Bernstein et al. 2001). Under natural conditions dsRNAs can be generated by transcription of transposons or viral sequences. An additional source of siRNAs might be antisense transcripts that are encoded on the opposite DNA strand of endogenous genes. Approximately 2500 antisense transcripts have been predicted for the human and mouse genomes, respectively (Lehner et al. 2003; Kiyosawa et al. 2003).

Though in principle siRNAs and also miRNAs are able to induce either RNAi or inhibition of translation (Doench et al. 2003; Zeng et al. 2003; Yekta et al. 2004), miRNAs are apparently more often involved in inhibition of translation than in RNAi, whereas siRNAs preferentially induce RNA degradation. One reason for this is probably the fact that siRNAs are complementary to their target sequence since they are usually transcribed from the antisense DNA strand of the target gene, whereas most miRNAs are generated from genes different from the target genes.

2
Posttranscriptional RNAi

The common RNAi pathway includes the processing of a dsRNA substrate into small dsRNAs by enzymes of the Dicer family, subsequent incorporation of the small dsRNAs into the RNA-induced silencing complex (RISC), and finally interaction of the miRNAs follow a similar pathway of processing and are apparently also incorporated into a RISC like complex that mediates repression of translation. As we show in this section, major, though not all, components of the involved protein machineries are conserved between different animal clades.

The formation of dsRNA substrates that can be processed by Dicer-like enzymes is assumed to occur in the nucleus, where the RNAs involved are synthesized. For miRNAs it has been shown that they are generated from longer transcripts, so-called pri-miRNAs, of which shorter approximately 70-nt-long hairpinlike precursors are excised in the nucleus (Lee et al. 2002, 2003). In *Drosophila melanogaster* and in human the RNAse III like enzyme, Drosha has been shown to mediate pri-miRNA cleavage (Han et al. 2004). The hairpin-like precursors termed pre-miRNAs are subsequently exported to the cytoplasm by Exportin 5 (Yi et al. 2003; Lund et al. 2004). In other organisms, little is known about formation and cellular localization of longer dsRNAs and miRNA precursors. In the cytoplasm, miRNA precursors are cleaved by Dicer, a second RNAse III like enzyme, into a small, imperfect dsRNA duplex with 2-nt-long 3′ overhangs; one of the two strands represents the 21–25-nt-long mature miRNA. Similarly, Dicer also cuts perfectly matching long dsRNAs into siRNAs. In *Caenorhabditis elegans* and in mammals only one Dicer enzyme has been identified, whereas *Drosophila* possesses two (Fig. 2). In *Caenorhabditis* processing of long dsRNAs requires interaction of Dicer1 with other proteins such as Rde-1, Rde-4 and an additional DEAD-box helicase, e.g., Drh-1 (Tabara et al. 2002). Rde-1 and Rde-4 appear to transport dsRNA to the Dicer complex, whereas the additional helicase Drh-1 might facilitate sliding of the Dicer complex on the dsRNA substrate. In *Caenorhabditis*, processing of miRNA precursors requires Alg-1 and Alg-2 that might also be involved in the transport of miRNA precursors to the Dicer complex (Grishok et al. 2001).

In *Drosophila*, Dicer1 preferentially processes pre-miRNAs, whereas Dicer2 appears to be involved only in the production of siRNAs (Lee et al.

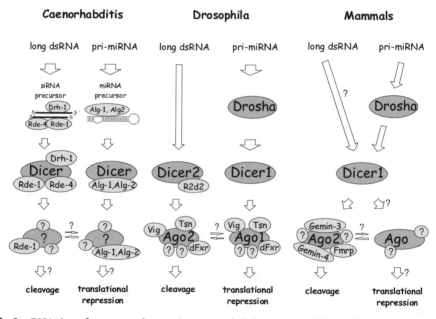

Fig. 2 RNA interference pathways in *Caenorhabditis*, *Drosophila* and mammals. The known protein components involved in miRNA and siRNA processing and function are shown. Cleavage of siRNA and miRNA precursors is mediated by Dicer-containing protein complexes (in *Drosophila* also by Drosha). In *Caenorhabditis*, miRNA and siRNA precursors are transported to the Dicer complexes by Alg-1/Alg-2 and Rde-1/Rde-4/Drh-1, respectively. RISC complexes contain Argonaute proteins (i.e. Ago1, Ago2) as key component and mediate the contact to the target mRNA. RISCs are responsible for cleavage of the mRNA or repression of protein synthesis. *dsRNA* double-stranded RNA

2004). During processing, Dicer2 forms a complex with the dsRNA binding protein R2d2 (Liu et al. 2003). This protein appears to stabilize the Dicer2/siRNA complex and to facilitate the subsequent contact to the RISC.

So far, in mammals only one Dicer enzyme, Dicer1, has been found (Hutvanger et al. 2001; Bernstein et al. 2003). Mammalian orthologs of Rde-1, Rde-4, Alg-1, Alg-2 and Dhr-1 have not been identified yet. Hence, it remains unclear if also in mammals processing and especially the transport of siRNA and miRNA precursors, respectively, require additional and possibly different proteins, thereby using special pathways.

Since components of the processing complex, such as Dicer1, Dicer2 and R2d2, interact with the RISC in *Drosophila* and *Caenorhabditis*, siRNAs and possibly also miRNAs are probably passed directly after cleavage to the RISC (Lee et al. 2004; Tabara et al. 2002). The RISC is a complicated multienzyme complex whose components have not all been identified (Fig. 2). The estimated sizes of the RISC range from 90 to 160 kDa in humans (Martinez et al. 2002) to 230–500 kDa in *Drosophila* (Hammond et al. 2001; Nykänen et al. 2001). The RISC contains an Argonaute protein as its signature component, which has two characteristic domains: PAZ, also found in Dicer family proteins, and PIWI (Carmell et al. 2002; Cerutti et al. 2000). The Argonaute protein family has multiple homologs in *Caenorhabditis* (24), *D. melanogaster* (five) and mammals (eight) (Carmell et al. 2002). Little is known about the components of the RISC in *Caenorhabditis*. Nevertheless, Rde-1 and also Alg-1 and Alg-2 contain PAZ and PIWI domains and are suggested to act in a RNA-induced-silencing-like complex in degradation and/or repression of translation of targeted mRNAs (Grishok et al. 2001; Tabara et al. 2002).

A recent study suggests that the RNA-induced silencing holo-complex in *Drosophila* is progressively assembled starting from an initiator complex that contains Dicer2 and R2d2 and the siRNA (Liu et al. 2003; Pham et al. 2004). The subsequent formation of the final 80S holo-complex requires Dicer1 and includes a number of intermediate complexes.

During this process the RISC undergoes an ATP-dependent activation step that results in unwinding of the double-stranded siRNAs delivered from Dicer (Nykänen et al. 2001), indicating that a RNA helicase activity is necessary for the functional RISC. The strand whose 5' end is more weakly bound to the complementary strand is thought to be stably attached to the PAZ domain of the Argonaute protein, whereas the other strand is expelled from the complex (Schwarz et al. 2003; Khvorova et al. 2003).

The final 80S holo-complex contains the Argonaute protein Ago2, and a number of different proteins, such as Vig, Tsn and dFxr (Fig. 2) (Pham et al. 2004). The existence of intermediate complexes might also provide an explanation for the observed different sizes of the RISC. For mammals major components of the RISC have been identified: Among these are the mammalian Ago1 (Elf2c1) and Ago2 (Elf2c2) proteins (Liu et al. 2004; Okumura et al. 2004; Martinez et al. 2002; Mourelatos et al. 2002; Carmell et al. 2002).

Additional components of the mammalian RISC are Gemin4 (Hhrf-1) and the DEAD-box RNA helicase Gemin3 (Ddx20) (Mourelatos et al. 2002; Hutvanger and Zamore 2002).

If the incorporated siRNA/miRNA shows sufficient complementarity to the target mRNA, the mRNA is cleaved (Hammond et al. 2000; Martinez et al. 2002; Elbashir et al. 2001). Recent experiments on Argonaute-containing complexes indicate that in *Drosophila* and mammals endonuclease activity might be restricted to Argonaute 2 proteins. (Meister et al. 2004; Liu et al. 2004). Cleavage proceeds via hydrolysis and the release of a 3′ hydroxyl and a 5′ phosphate terminus.

In contrast to mRNA cleavage, inhibition of translation requires only partial complementarity of mRNA and miRNA. In mice, inhibition of protein synthesis does not seem to require Ago2 (Liu et al. 2004). In *Drosophila*, mRNA cleavage appears to depend on Ago2, whereas for inhibition of translation Ago1 seems to be necessary (Okamura et al. 2004). A second protein which is apparently necessary for the miRNA-induced inhibition of translation is the human fragile X protein FMRP (FMR-1), and its *Drosophila* ortholog dFxr, respectively. In both organisms, this protein interacts with components of the RISC (Ishizuka et al. 2002; Caudy et al. 2002; Jin et al. 2004). Biochemical studies on fragile X protein suggest a role in inhibition of translation (Laggerbauer et al. 2000; Schaeffer et al. 2001). Interestingly, reduction in Ago1 expression rescues the rough-eye phenotype of dFxr over-expressing *Drosophila* mutants, indicating Ago1 is required for dFxr function (Jin et al. 2004). Fragile X protein deficiency is related to specific neuronal defects and seems to function as a component in a cell-type-specific Ago1/RISC. In conclusion, in animals several differently composed silencing complexes might exist that trigger either target mRNA cleavage or inhibition of protein synthesis.

In plants, fungi and worms, dsRNAs can be newly synthesized by RNA-dependent RNA polymerases using short RNAs as primers (Nishikura 2001). This results in amplification of dsRNAs that serve as substrates for RNAi. Thus, starting from small amounts of dsRNAs, these enzymes are thereby able to enhance the RNAi. So far, a gene that might encode a mammalian ortholog of these RNA-dependent RNA polymerases has not been identified in the human and mouse genomes. Hence, a similar amplification mechanism might not be present in mammals, implying that not all mechanisms of RNAi might be conserved among metazoans.

3
Developmental Gene Regulation by miRNA-Mediated RNAi

In mice, the function of the RNAi machinery appears to be essential for embryonic development since homozygote loss of function mutations in Dicer1 and Ago2, respectively, result in lethality during early embryonic stages (Bernstein et al. 2003; Liu et al. 2004). Dicer-deficient embryos die around day 8.5 of embryonic development. Vital Ago2-deficient embryos can still be found at day 10.5, though at this time they already exhibit severe developmental defects. Dicer1 deficiency might result in a complete breakdown of the dsRNA processing machinery and subsequent mRNA degradation and inhibition of translation. Ago2 deficiency might only affect the mRNA degradation pathway which might be the reason for the longer lifespan of affected embryos compared with Dicer-deficient embryos.

Tissue-specific expression of mammalian miRNAs indicates that the function of individual miRNAs is restricted to distinct tissues. Sempere et al. (2004) found that out of 119 miRNAs analyzed 66 are expressed in adult brain or in brain development. Further miRNAs were enriched in lung (ten), spleen (seven), hematopoietic tissues (three), liver (five), heart (six) and kidney (nine). In this study, only nine miRNAs analyzed were ubiquitously expressed. The large portion of miRNAs expressed in brain might indicate that they play a major role in neuronal development and differentiation. miRNAs also seem to contribute to physiological changes under pathological conditions since *miR-26a* and *miR-99a* expression is reduced in human lung cancer cells (Calin et al. 2004) and the *miR-155* precursor is enriched in Burkitt lymphoma tissues (Metzler et al. 2004). Thus, overexpression or nonexpression of miRNAs might contribute to the development or maintenance of cells to be transformed to cancer cells.

So far little is known about the mammalian genes that are specifically regulated by miRNAs. In contrast, target genes of some miRNAs in *Caenorhabditis* and *Drosophila* have been identified and the phenotypical functions of these miRNAs have been investigated in detail. Among these miRNAs is the *let-7* miRNA in *Caenorhabditis* that emerged from mutational screens for genes regulating larval development (Reinhart et al. 2000). Interestingly, this miRNA appears to be present also in fly and in vertebrate genomes. Thus, it could be hypothesized that the function of *let-7* might also be conserved in these clades. In *Caenorhabditis*, *let-7* is expressed during late larval and adult stages and regulates the *lin-41* and *hbl-1* genes (Slack et al. 2000; Abrahante et al. 2003; Lin et al. 2003). Loss of *let-7* activity results in reiteration of larval cell fates during the adult stage, whereas an elevated *let-7* dose causes precocious expression of adult fates during larval stages (Reinhart et al. 2000). Whether the mammalian *let-7* miRNAs interfere with mammalian homologs of the *Caenorhabditis* target genes that act in analogous pathways during mammalian development is so far unclear.

A recent study (Yekta et al. 2004) indicates that subtypes of the miRNA *miR-196* might contribute to the regulation of mammalian development by interaction with *Hox* genes. Slightly different variants of *miR-196* are encoded in similar positions in the three mammalian *Hox* clusters *Hoxa*, *Hoxb* and *Hoxc*. Transcripts of four mammalian *Hox* genes, *Hoxa7*, *Hoxb8*, *Hoxc8* and *Hoxd8*, possess conserved sequence motifs that are very similar to the complementary *miR-196a* sequence. In mouse embryos *miR-196a* appears to induce cleavage of *Hoxb8* transcripts that possess a complementary target motif containing only one mismatch. In cell culture assays *miR-196a* also induced degradation of reporter gene transcripts that harbored the same target motif. In addition, motifs with more mismatches in the 3' untranslated regions of *Hoxa7*, *Hoxc8* and *Hoxd8* were sufficient to reduce reporter gene expression most likely by inhibition of translation. Thus, *miR-196a* might be able to regulate the expression of a number of *Hox* genes by different mechanisms.

In a recent publication it was shown that *miR-181* is highly expressed in bone-marrow B cells and in thymus (Chen et al. 2004). Overexpression of this miRNA in bone-marrow hematopoietic progenitor cells results in higher numbers of B cells and a decrease in the numbers of CD8+ T cells. Though *miR-181* loss of function mutations have not yet been established the data obtained so far suggest that *miR-181* functions in induction of B cell differentiation.

4
Epigenetic Gene Silencing

In vertebrate genomes one major aspect of gene regulation is represented by changes in chromatin conformation, which influences the accessibility of the DNA for the transcriptional machinery. This is mostly mediated by epigenetic modifications of DNA and histones. Epigenetic modifications can be, on one hand, transmitted during cell division onto the daughter cells and, on the other hand, they are reversible. This suggests that the genome-wide patterns of epigenetic modifications are cell-type-specific and are submitted to changes during development of the organism. Cytosine methylation of CpG dinucleotides and acetylation or methylation of the histones H3 and H4 represent the major epigenetic modifications in vertebrates. Specific epigenetic modifications in the promoter regions of a large number of genes have been described as being crucial for gene expression. In most cases, methylation of CpGs and certain types of histone modifications are associated with a compact, transcription-repressing chromatin structure, whereas histone acetylation is related to an open, transcription-inducing chromatin structure (Jenuwein and Allis 2001; Turner 2002).

Besides its function in regulation of individual genes, epigenetic gene regulation has been shown to be crucial for chromosome-wide silencing of genes on the inactive X chromosome in female mammals, and also for the regulation of genomic imprinting in mammals (Reik and Walter 2001; Walter and Paulsen 2003; Anderson and Panning 2003). Imprinted genes are preferentially expressed from only one of the two parental alleles, and the parental origin of the allele determines if it is expressed or silenced. The frequent appearance of noncoding RNAs in imprinting domains and the involvement of noncoding RNAs in X chromosome inactivation has led to the assumption that noncoding RNAs might play an important role in epigenetic gene silencing.

5
Small dsRNAs May Mediate Epigenetic Silencing of Genes

Though not yet analyzed in detail, there are many lines of evidence that dsRNAs are involved in epigenetic processes such as gene silencing and centromere stabilization in vertebrates. Fukagawa et al. (2004) have shown that Dicer deficiency led to centromere instability in chicken–human hybrid cell lines that contained the human chromosome 21 in a chicken background. Absence of Dicer resulted in chromosomal missegration caused by premature sister chromatid separation. This was associated with an aberrant chromatin indicated by the absence of normal HP1 protein binding to centromeres, especially during mitosis. In addition, aberrant accumulation of large amounts of overlapping long sense and antisense transcripts in human chromosome 21 α-satellite sequences was observed. Whereas in presence of Dicer low amounts of small RNAs derived from the satellite sequences were detected, these were almost completely absent in Dicer-deficient cells. This suggests that this protein mediates degradation of centromeric dsRNAs under normal conditions. Although a direct relationship between expression of small dsRNAs and centromere stability was not shown, the observed effects are reminiscent to the transcription of centromere repeats in *Schizosaccharomyces pombe* (Volpe et al. 2002).

Interestingly, dsRNAs might also be involved in induction of gene expression as is indicated for neuron-specific genes that are regulated by the neuronal restricted silencing factor NRSF/REST. Differentiation of mammalian adult neural stem cells into neurons is mediated by induction of neuron-specific genes that are regulated by NRSF/REST. This factor represses these genes in silent stem cells by binding to the promoter regions via a conserved binding site. Kuwabara et al. (2004) identified a small dsRNAs of unknown genomic origin that contained a 21-nt-long binding motif for NSRF/REST in hippocampal neural stem cells. Expression of these small dsRNAs resulted in induction of genes that are regulated by NRSF/REST and subsequently to differentiation into neurons. This was related to changes of histone modifications and chromatin-associated proteins in the promoter regions of affected genes, indicating a direct influence of the dsRNAs on chromatin structure. Whereas in centromere regions the RNAi machinery appears to induce a heterochromatin structure, expression of NRSF/REST binding dsRNAs is related to an open chromatin structure, indicating that two different mechanisms are involved. Preferential binding of NRSF/REST to the dsRNAs rather than to DNAs suggests that the binding to dsRNA inhibits the attachment of the protein factor to the promoter region.

Though these publications provide evidence for small RNA mediated changes in chromatin structure and DNA methylation in amniotes the mechanisms involved are poorly understood. As described before, the necessary small dsRNAs appear to be processed by Dicer. This enzyme is usually located in the cytoplasm; thus, transport of dsRNAs back into the nucleus appears to be necessary as indicated by the centromere destabilization in the case of Dicer deficiency. Alternatively, a nuclear Dicer-like enzyme might exist.

So far, dsRNA-mediated changes in epigenetic gene regulation have only been shown in mammalian cell culture systems but not under natural conditions in mammalian organisms. In a recent study (Svoboda et al. 2004) expression of a long dsRNA targeted against the *Mos* gene led to RNAi-mediated silencing of the *Mos* gene but failed to induce DNA methylation in oocytes of transgenic mice. However, this does not necessarily exclude that in mammals DNA methylation might be induced by dsRNAs since such processes might be restricted to specific developmental stages or to distinct tissues. Secondly, genomic DNA can be protected against de novo methylation by DNA bind-

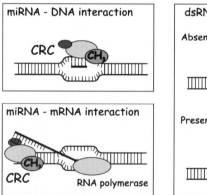 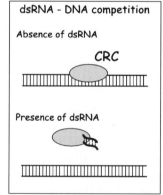

Fig. 3 Models for interaction of short RNAs with chromatin remodeling complexes (*CRCs*). RNA–DNA interaction: Short RNAs might hybridize with complementary DNA motifs. Such distortions in the DNA structure might be recognized by CRCs that target chromatin-modifying enzymes such as DNA methyltransferases (CH_3) to the DNA. RNA–mRNA interaction: Short RNAs might hybridize to nascent precursor mRNAs. CRCs could bind to these duplexes. Since nascent RNAs are still attached to the DNA, the CRCs get into close contact with the chromatin. dsRNA–DNA competition: dsRNAs might compete with DNA in binding to chromatin-remodeling complexes or mediator proteins, transcription factors or other DNA binding proteins. Removal of these proteins from the binding sites could make the DNA accessible for changes in chromatin structure

ing proteins as has been shown for the allele-specifically methylated region upstream of *H19* which becomes de novo methylated in oocytes in the absence of the DNA binding protein CTCF (Fedoriw et al. 2004). Thus, a distinct chromatin structure might make the DNA inaccessible to dsRNA-mediated epigenetic changes.

Though there is obviously strong evidence for a role of short RNAs in epigenetic gene silencing the mechanisms involved are poorly understood. It has been assumed that RNAs target chromatin remodeling complexes (CRCs) to specific sites in the genome (Grewal and Moazed 2003). For this, different possible mechanistic models can be postulated (Fig. 3): (1) CRCs might recognize DNA–RNA hybrids; (2) during transcription short RNAs might interact with nascent transcripts, thereby mediating close contact of RNA–RNA binding CRCs to the DNA (Grewal and Moazed 2003); (3) the study on NRSF/REST mediated gene silencing indicates that binding to dsRNAs and DNA might prevent binding of repressor proteins to promoter regions (Kuwabara et al. 2004).

6
Epigenetic Silencing by Antisense Transcripts

Inspired by studies on epigenetic changes in tumor suppressor genes during tumorgenesis, it has been suggested that antisense transcription might be a possible cause of epigenetic silencing of these genes (Jones 1999). Aberrant activation of intronic promoter-like elements, for example, repetitive *Alu* sequences, might lead to antisense transcription into the promoter regions of tumor suppressor genes. As a result the promoter region might get hypermethylated and the tumor suppressor gene silenced (Fig. 4).

That such silencing effects might indeed be possible is indicated by a study on α-thalassemia patients who harbor small deletions downstream of the hemoglobin α-2 gene (Tufarelli et al. 2003). In mutant mice that harbor a similar mutation expanded transcription of a neighboring gene and thereby antisense transcription of the hemoglobin α-2 gene have been observed. The antisense transcript overlaps with a CpG island at the promoter of the hemoglobin α-2 gene. Associated with this, hypermethylation of the CpG island and hemoglobin α-2 repression were detected, suggesting that the antisense transcript might trigger DNA methylation and gene silencing.

Compared with nonimprinted genes, a rather large number of imprinted genes have been predicted to overlap with antisense transcripts in mice and humans (Kiyosawa et al. 2003; Lehner et al. 2002). Antisense transcription has also been shown experimentally for a number of imprinted genes (summarized in Reik and Walter 2001). The most prominent imprinted antisense transcript is the paternally expressed transcript *Air* that overlaps with the

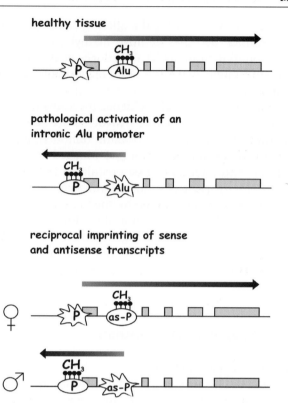

Fig. 4 RNA antisense transcription induces DNA methylation. In healthy tissues, methylation of intronic repetitive elements such as *Alu* elements prevents aberrant antisense transcription. In pathological situations these elements become demethylated and aberrant antisense transcription is initiated. This results in methylation and repression of the promoter. In the case of imprinted genes similar effects are seen under natural conditions on the different parental alleles. Allele-specific methylation of sense (*P*) and anti-sense (*as-P*) promoters results in allele-specific transcription of the gene and its antisense transcript

maternally expressed *Igf2r* transcript in mice (Fig. 4). *Igf2r* contains a CpG island in its second intron that is only methylated on the maternal allele and serves apparently as a promoter of the paternally expressed antisense transcript *Air*. In mutant mice introduction of a polyadenylation signal in the 5′ portion of the *Air* transcript leads to premature termination of the transcript and subsequently to biallelic expression of the oppositely orientated *Igf2r* gene (Sleutels et al. 2002). Interestingly, the neighboring *Slc22a2* and *Slc22a3* genes that do not overlap with *Air* show loss of imprinting, suggesting that in this case gene silencing is not only based on transcription in antisense orientation but also on additional so far unknown chromatin remodeling effects of the *Air* RNA.

7
Dose Compensation on Sex Chromosomes

Noncoding RNAs are also involved in gene dose compensation on sex chromosomes as shown for the mammalian *Xist* RNA and the *roX* RNAs in *Drosophila*. In humans one female X chromosome is inactivated, whereas in *Drosophila* the single male X chromosome is activated twofold. In both cases a modulation of chromatin organization of the relevant genes leads to changes of the transcription rates. *Xist* and *roX* RNAs are examples for the fact that noncoding RNAs are able to stimulate transcription as well as to silence gene expression. In the following the influence of noncoding RNAs on dose compensation in flies and mammals is discussed in more detail.

8
The Role of *roX* RNAs in Dose Compensation in *Drosophila*

Dose compensation in *Drosophila* is mediated by a ribonucleoprotein (RNP) complex which forms in male cells and coats most of the X chromosome in *cis*. The coating initiates at 30–40 chromatin entry sites and results in bidirectional spreading of hyperactive chromatin and thereby in hypertranscription of underlying genes (Kelley et al. 1999).

This ribonucleoprotein male-specific lethal (MSL) complex consists of six proteins (MSL1-3, MLE, MOF, JIL1) and the two noncoding RNAs *roX1* and *roX2* that are encoded on the X chromosome. The functions of these RNAs in the MSL complex are not yet completely understood. Although both RNAs show little sequence homology and are with 3.7 kb for *roX1* and 0.5–2 kb for *roX2* also different in size (Amrein and Axel 1997; Smith et al. 2000) they are functionally redundant. Deletion of either *roX* gene has no effect on males, whereas losing both of them results in male lethality (Meller and Rattner 2002; Franke and Baker 1999). Further deletion studies identified a functionally important long double-stranded stem-loop structure within the 3' end of *roX1* (Stuckenholz et al. 2003). *RoX* RNAs are stabilized by the MSL complex and are also required for the assembly of this complex at the chromatin entry sites. *RoX* RNAs are also necessary for the spreading of the complex along the X chromosome and for proper dose compensation in general (Park et al. 2002; Andersen and Panning 2003). A role of *roX* RNA in targeting the MSL complex specifically to the X chromosome has also been suggested (Meller et al. 2000; Park et al. 2002, 2003). At the moment it remains unclear if the entry sites are recognized by specific binding of *roX* RNAs or proteins of the MSL complex to chromosomal DNA as proposed by Kageyama et al. (2001). A second hypothesis suggests that starting at the sites of their synthesis *roX* RNAs may mark the X chromosome for dose compensation. The RNA coat may subsequently attract the MSL proteins specifically to the respective chromosome (Kelley and Kuroda 2000). A feature which clearly distinguishes the *roX* RNAs from their mammalian relative *Xist* is that they are able to act in *trans*. This was shown by several groups which observed that a MSL complex containing *roX* RNA expressed from an autosomal transgene was able to spread in *cis* along the autosome as well as in *trans* along the X chromosome (Kelley et al. 1999; Henry et al. 2001).

9
X Chromosome Inactivation in Mammals

The process of X chromosome inactivation in mammals is a multistep process involving chromosomal RNA coating, and several chromatin modifications, including hypoacetylation and methylation of histone H3, methylation of CpGs and hypoacetylation of histone H4 (Plath et al. 2003). The inactivation is initiated by the stabilization of a noncoding RNA, termed *Xist* (X inactive-specific transcript) which is transcribed from the X inactivation center (Xic) on the X chromosome to be inactivated. The *Xist* RNA spreads in *cis* over 100 Mb from its site of synthesis, thereby coating the X chromosome. This is followed by heterochromatin formation and transcriptional silencing of the *Xist*-expressing X chromosome (reviewed in Cohen and Lee 2002; Wutz 2003). Once established, silencing of the inactivated X chromosome is stable and is maintained clonally even if *Xist* is removed (Wutz and Jaenisch 2000; Brown and Willard 1994; Csankowski et al. 1999; Hall et al. 2002).

The *Xist* RNA contains domains responsible for silencing and coating of the X chromosome. Conserved sequences in the 5' end of the RNA in all species analyzed to date suggest the importance of this region for transcriptional silencing. The respective region comprises 7.5 repeat units termed A repeats that are predicted to form RNA stem-loop structures. The deletion of this region results in constantly efficient coating but loss of the ability to silence the chromosome (Fig. 5) (Wutz et al. 2002). For coating of the X chromosome the 3' region of *Xist* is required. Deletion or blockage of the 3' end abolishes both coating and silencing (Wutz et al. 2002; Beletskii et al. 2001).

Changes in epigenetic modifications on the inactivated X chromosome follow immediately the spreading of *Xist* RNA and seem to coincide with transcriptional silencing (Heard et al. 2001; Mermoud et al. 2002). This raises the question whether recruitment of a protein complex with deacetylase and/or methyltransferase activity may be one function of *Xist* RNA or whether *Xist* may be part of an RNP chromatin remodeling complex comparable with the MSL complex in *Drosophila* (reviewed in Andersen and Panning 2003). Of particular interest in this context is the human tumor suppressor gene *BRCA1*, which represents the first possible member of such a complex. The human BRCA1 protein has been shown to colocalize with *XIST* during the

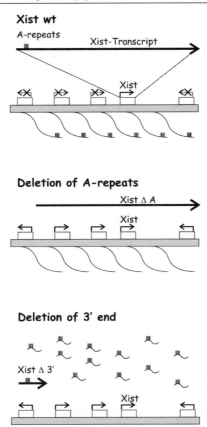

Fig. 5 *Xist*-mediated X chromosome inactivation. Transcription of *Xist* results in coating of the X chromosome in *cis* and silencing of genes on the inactive X chromosome. Truncated *Xist* transcripts lacking the A repeats are still able to coat the chromosome but not to silence genes. *Xist* transcripts in which the 3′ region is deleted are not able to coat the chromosome and to silence genes

S-phase in many female somatic cells and seems to play a role in maintenance of random X inactivation. In humans, loss of BRCA1 results in defective X inactivation through failure of *XIST* RNA to coat the X chromosome and the loss of H3K9 methylation and of macrohistone variant H2A1 association. The reconstitution with wild-type BRCA1 reestablishes *XIST* RNA coating (Ganesan et al. 2002). Interestingly BRCA1 interacts directly via its ring finger domain with *XIST* and with RNA helicase A, the human homolog of *Drosophila* MLE which is required for chromosome coating by the MSL complex (Anderson et al. 1998; Lee and Hurwitz 1993). This clearly strengthens the hypothetical existence of a chromatin remodeling complex.

In early cleavage stages and extraembryonic tissues of later stage preferential inactivation of the paternal X chromosome has been observed in mice

(Takagi and Sasaki 1975). The role of epigenetic modifications, e.g., DNA methylation or histone modifications in imprinted X chromosome inactivation, are still unclear. During later cleavage stages X chromosome inactivation changes to a random process in the embryo proper, i.e., maternally and paternally derived X chromosomes exhibit equal probability of being inactivated in every cell.

Lee et al. (1999) identified *Tsix* as an antisense transcript of *Xist* that negatively regulates *Xist* by decreasing the steady-state levels of *Xist* RNA. *Tsix* seems to play a crucial role in random X chromosome choice (Lee and Lu 1999; Lee 2000). Deletion of *Tsix* *cis* regulatory elements or insertion of a polyadenylation signal cassette downstream of *Tsix* has been shown to abolish stochastic X chromosome choice in differentiating stem cells. In most cases the mutant X chromosome became the inactive X (Lee and Lu 1999; Luikenhuis et al. 2001; Stavropoulos et al. 2001; Sado et al. 2001).

Loss of *Tsix* generally can result in a slight upregulation of *Xist* (Lee and Lu 1999). The molecular mechanisms of *Xist* regulation by *Tsix* are not known so far but several theories have been proposed (Andersen and Panning 2003): Transcription of *Tsix* might decrease the transcription of *Xist*, or *Tsix* RNA could destabilize *Xist* RNA through posttranscriptional mechanisms such as RNAi. At the moment the favored hypothesis postulates that a *Tsix–Xist* RNA duplex may interfere with the structure of *Xist* RNA, thereby preventing the formation of a functional *Xist* RNP. Absence of the complex could lead to reduced stability of the *Xist* RNA as observed for *roX* RNAs in *Drosophila*. Another hypothesis is supported by the finding that *Tsix* RNA and all of its identified spliced isoforms overlap the A repeats of *Xist* RNA (Sado et al. 2001; Shibata and Lee 2003) and thus may prevent correct folding and function of the *Xist* RNA silencing domain.

10
Conclusions

Similar to the situation in plants, RNAs appear to be involved in a number of gene silencing mechanisms in animals such as RNAi and chromatin-associated epigenetic silencing. The observed conservation of enzymes, structural similarities of the RNAs involved, and induction of similar epigenetic modifications suggest that the pathways of RNA-mediated silencing are highly conserved in metazoans. Nevertheless, some components such as RNA-dependent RNA polymerases might be absent in mammals, whereas other mechanisms such as RNA-mediated dose compensation appear to be present only in animal species but have not yet been observed in plants.

Surprisingly small RNA induced pathways for different silencing mechanisms appear to overlap: RNAi and repression of translation can be mediated by miRNAs and siRNAs, and also the protein complexes involved are apparently shared among both pathways. Similarly, epigenetic silencing can be mediated by short RNAs, suggesting that at least for some steps of the process the same or a very similar protein machinery is used. A major function of short RNAs and also longer RNAs such as *Xist* might be the recruitment of histone deacetylases and DNA methylases to specific sites in the genome such as promoter regions and sex chromosomes though the mechanisms involved appear to be different for short and long RNAs, respectively. Although many details of the functions and targets of noncoding RNAs are still enigmatic, it is clear that the functional analysis of genomes and epigenomes will require more focus on RNA-mediated mechanisms.

Acknowledgements We thank the Deutsche Forschungsgemeinschaft for support by grant WA1029/3-1.

References

Abrahante JE, Daul AL, Li M, Volk ML, Tennessen JM, Miller EA, Rougvie AE (2003) The Caenorhabditis elegans hunchback-like gene lin-57/hbl-1 controls developmental time and is regulated by microRNAs. Dev Cell 4:625–637

Amrein H, Axel R (1997) Genes expressed in neurons of adult male Drosophila. Cell 88:459–469

Andersen AA, Panning B (2003) Epigenetic gene regulation by noncoding RNAs. Curr Opin Cell Biol 15:281–289

Anderson SF, Schlegel BP, Nakajima T, Wolpin ES, Parvin JD (1998) BRCA1 protein is linked to the RNA polymerase II holoenzyme complex via RNA helicase A. Nat Genet 19:254–256

Baulcombe D (1999) Viruses and gene silencing in plants. Arch Virol Suppl 15:189–201

Beletskii A, Hong YK, Pehrson J, Egholm M, Strauss WM (2001) PNA interference mapping demonstrates functional domains in the noncoding RNA Xist. Proc Nat Acad Sci USA 98:9215–9220

Bernstein E, Caudy AA, Hammond SM, Hannon GJ (2001) Role for a bidentate ribonuclease in the initiation step of RNA interference. Nature 409:363–366

Bernstein E, Kim SY, Carmell MA, Murchison EP, Alcorn H, Li MZ, Mills AA, Elledge SJ, Anderson KV, Hannon GJ (2003) Dicer is essential for mouse development. Nat Genet 35:215–217

Brown CJ, Willard HF (1994) The human X-inactivation centre is not required for maintenance of X-chromosome inactivation. Nature 368:154–156

Calin GA, Sevignani C, Dumitru CD, Hyslop T, Noch E, Yendamuri S, Shimizu M, Rattan S, Bullrich F, Negrini M, Croce CM (2004) Human microRNA genes are frequently located at fragile sites and genomic regions involved in cancers. Proc Natl Acad of Sci USA 101:2999–3004

Carmell MA, Xuan Z, Zhang MQ, Hannon GJ (2002) The Argonaute family: tentacles that reach into RNAi, developmental control, stem cell maintenance, and tumorigenesis. Genes Dev 16:2733–2742

Caudy AA, Myers M, Hannon GJ, Hammond SM (2002) Fragile X-related protein and VIG associate with the RNA interference machinery. Genes Dev 16:2491–2496

Cerutti L, Mian N, Bateman A (2000) Domains in gene silencing and cell differentiation proteins: the novel PAZ domain and redefinition of the Piwi domain. Trends Biochem Sci 25:481–482

Chen CZ, Li L, Lodish HF, Bartel DP (2004) MicroRNAs modulate hematopoietic lineage differentiation. Science 303:83–86

Cohen DE, Lee JT (2002) X-chromosome inactivation and the search for chromosome-wide silencers. Curr Opin Genet Dev 12:219–224

Csankovszki G, Panning B, Bates B, Pehrson JR, Jaenisch R (1999) Conditional deletion of Xist disrupts histone macroH2A localization but not maintenance of X inactivation. Nat Genet 22:323–324

Debrand E, Chureau C, Arnaud D, Avner P, Heard E (1999) Functional analysis of the DX-Pas34 locus, a 3' regulator of Xist expression. Mol Cell Biol 19:8513–8525

Doench JG, Petersen CP, Sharp PA (2003) siRNAs can function as miRNAs. Genes Dev 17:438–442

Dostie J, Mourelatos Z, Yang M, Sharma A, Dreyfuss G (2003) Numerous microRNPs in neuronal cells containing novel microRNAs. RNA 9:180–186

Elbashir SM, Harborth J, Lendeckel W, Yalcin A, Weber K, Tuschl T (2001) Duplexes of 21-nucleotide RNAs mediate RNA interference in cultured mammalian cells. Nature 411:494–498

Fedoriw AM, Stein P, Svoboda P, Schultz RM, Bartolomei MS (2004) Transgenic RNAi reveals essential function for CTCF in H19 gene imprinting. Science 303:238–240

Fire A, Xu S, Montgomery MK, Kostas SA, Driver SE, Mello CC (1998) Potent and specific genetic interference by double-stranded RNA in Caenorhabditis elegans. Nature 391:806–811

Franke A, Baker BS (1999) The rox1 and rox2 RNAs are essential components of the compensasome, which mediates dosage compensation in *Drosophila*. Mol Cell 4:117–122

Fukagawa T, Nogami M, Yoshikawa M, Ikeno M, Okazaki T, Takami Y, Nakayama T, Oshimura M (2004) Dicer is essential for formation of the heterochromatin structure in vertebrate cells. Nat Cell Biol 6:784–791

Ganesan S, Silver DP, Greenberg RA, Avni D, Drapkin R, Miron A, Mok SC, Randrianarison V, Brodie S, Salstrom J, Rasmussen TP, Klimke A et al. (2002) BRCA1 supports XIST RNA concentration on the inactive X chromosome. Cell 111:393–405

Grewal SI, Moazed D (2003) Heterochromatin and epigenetic control of gene expression. Science 301:798–802

Grishok A, Pasquinelli AE, Conte D, Li N, Parrish S, Ha I, Baillie DL, Fire A, Ruvkun G, Mello CC (2001) Genes and mechanisms related to RNA interference regulate expression of the small temporal RNAs that control *Caenorhabditis* developmental timing. Cell 106:23–34

Hall LL, Byron M, Sakai K, Carrel L, Willard HF, Lawrence JB (2002) An ectopic human XIST gene can induce chromosome inactivation in postdifferentiation human HT-1080 cells. Proc Natl Acad Sci USA 99:8677–8682

Hammond SM, Bernstein E, Beach D, Hannon GJ (2000) An RNA-directed nuclease mediates post-transcriptional gene silencing in *Drosophila* cells. Nature 404:293–296

Hammond SM, Boettcher S, Caudy AA, Kobayashi R, Hannon GJ (2001) Argonaute2, a link between genetic and biochemical analyses of RNAi. Science 293:1146–1150

Han J, Lee Y, Yeom KH, Kim YK, Jin H, Kim VN (2004) The Drosha-DGCR8 complex in primary microRNA processing. Genes Dev 18:3016–3027

Heard E, Rougeulle C, Arnaud D, Avner P, Allis CD, Spector DL (2001) Methylation of histone H3 at Lys-9 is an early mark on the X chromosome during X inactivation. Cell 107:727–738

Henry RA, Tews B, Li X, Scott MJ (2001) Recruitment of the male-specific lethal (MSL) dosage compensation complex to an autosomally integrated roX chromatin entry site correlates with an increased expression of an adjacent reporter gene in male *Drosophila*. J Biol Chem 276:31 953–31 958

Hutvagner G, McLachlan J, Pasquinelli AE, Balint E, Tuschl T, Zamore PD (2001) A cellular function for the RNA-interference enzyme Dicer in the maturation of the let-7 small temporal RNA. Science 293:834–838

Hutvagner G, Zamore PD (2002) A microRNA in a multiple-turnover RNAi enzyme complex. Science 297:2056–2060

Ishizuka A, Siomi MC, Siomi H (2002) A *Drosophila* fragile X protein interacts with components of RNAi and ribosomal proteins. Genes Dev 16:2497–2508

Jenuwein T, Allis CD (2001) Translating the histone code. Science 293:1074–1080

Jin P, Zarnescu DC, Ceman S, Nakamoto M, Mowrey J, Jongens TA, Nelson DL, Moses K, Warren ST (2004) Biochemical and genetic interaction between the fragile X mental retardation protein and the microRNA pathway. Nat Neurosci 7:113–117

Jones PA (1999) The DNA methylation paradox. Trends Genet 15:34–37

Kageyama Y, Mengus G, Gilfillan G, Kennedy HG, Stuckenholz C, Kelley RL, Becker PB, Kuroda MI (2001) Association and spreading of the *Drosophila* dosage compensation complex from a discrete roX1 chromatin entry site. EMBO J 20:2236–2245

Kelley RL, Meller VH, Gordadze PR, Roman G, Davis RL, Kuroda MI (1999) Epigenetic spreading of the *Drosophila* dosage compensation complex from roX RNA genes into flanking chromatin. Cell 98:513–522

Kelley RL, Kuroda MI (2000) The role of chromosomal RNAs in marking the X for dosage compensation. Curr Opin Genet Dev 10:555–561

Khvorova A, Reynolds A, Jayasena SD (2003) Functional siRNAs and miRNAs exhibit strand bias. Cell 115:209–216

Kiyosawa H, Yamanaka I, Osato N, Kondo S, Hayashizaki Y; RIKEN GER Group; GSL Members (2003) Antisense transcripts with FANTOM2 clone set and their implications for gene regulation. Genome Res 13:1324–1334

Kuwabara T, Hsieh J, Nakashima K, Taira K, Gage FH (2004) A small modulatory dsRNA specifies the fate of adult neural stem cells. Cell 116:779–793

Laggerbauer B, Ostareck D, Keidel EM, Ostareck-Lederer A, Fischer U (2001) Evidence that fragile X mental retardation protein is a negative regulator of translation. Hum Mol Genet 10:329–338

Lagos-Quintana M, Rauhut R, Lendeckel W, Tuschl T (2001) Identification of novel genes coding for small expressed RNAs. Science 294:853–858

Lagos-Quintana M, Rauhut R, Yalcin A, Meyer J, Lendeckel W, Tuschl T (2002) Identification of tissue-specific microRNAs from mouse. Curr Biol 12:735–739

Lee CG, Hurwitz J (1993) Human RNA helicase A is homologous to the maleless protein of *Drosophila*. J Biol Chem 268:16 822–830

Lee JT, Lu N (1999) Targeted mutagenesis of Tsix leads to nonrandom X inactivation. Cell 99:47–57

Lee JT, Davidow LS, Warshawsky D (1999) Tsix, a gene antisense to Xist at the X-inactivation centre. Nat Genet 21:400–404

Lee Y, Jeon K, Lee JT, Kim S, Kim VN (2002) MicroRNA maturation: stepwise processing and subcellular localization. EMBO J 21:4663–4670

Lee Y, Ahn C, Han J, Choi H, Kim J, Yim J, Lee J, Provost P, Radmark O, Kim S, Kim VN (2003) The nuclear RNase III Drosha initiates microRNA processing. Nature 425:415–419

Lee YS, Nakahara K, Pham JW, Kim K, He Z, Sontheimer EJ, Carthew RW (2004) Distinct roles for *Drosophila* Dicer-1 and Dicer-2 in the siRNA/miRNA silencing pathways. Cell 117:69–81

Lehner B, Williams G, Campbell RD, Sanderson CM (2002) Antisense transcripts in the human genome. Trends Genet 18:63–65

Lin SY, Johnson SM, Abraham M, Vella MC, Pasquinelli A, Gamberi C, Gottlieb E, Slack FJ (2003) The C. elegans hunchback homolog, hbl-1, controls temporal patterning and is a probable microRNA target. Dev Cell 4:639–650

Liu Q, Rand TA, Kalidas S, Du F, Kim HE, Smith DP, Wang X (2003) R2D2, a bridge between the initiation and effector steps of the *Drosophila* RNAi pathway. Science 301:1921–1925

Liu J, Carmell MA, Rivas FV, Marsden CG, Thomson JM, Song JJ, Hammond SM, Joshua-Tor L, Hannon GJ (2004) Argonaute2 is the catalytic engine of mammalian RNAi. Science 305:1437–1441

Luikenhuis S, Wutz A, Jaenisch R (2001) Antisense transcription through the Xist locus mediates Tsix function in embryonic stem cells. Mol Cell Biol 21:8512–8520

Lund E, Guttinger S, Calado A, Dahlberg JE, Kutay U (2004) Nuclear export of microRNA precursors. Science 303:95–98

Martinez J, Patkaniowska A, Urlaub H, Luhrmann R, Tuschl T (2002) Single-stranded antisense siRNAs guide target RNA cleavage in RNAi. Cell 110:563–574

Meister G, Landthaler M, Patkaniowska A, Dorsett Y, Teng G, Tuschl T (2004) Human Argonaute2 mediates RNA cleavage targeted by miRNAs and siRNAs. Mol Cell 15:185–197

Meller VH, Rattner BP (2002). The roX genes encode redundant male-specific lethal transcripts required for targeting of the MSL complex. EMBO J 21:1084–1091

Meller VH, Gordadze PR, Park Y, Chu X, Stuckenholz C, Kelley RL, Kuroda MI (2000) Ordered assembly of roX RNAs into MSL complexes on the dosage-compensated X chromosome in *Drosophila*. Curr Biol 10:136–143

Mermoud JE, Popova B, Peters AH, Jenuwein T, Brockdorff N (2002) Histone H3 lysine 9 methylation occurs rapidly at the onset of random X chromosome inactivation. Curr Biol 12:247–251

Metzler M, Wilda M, Busch K, Viehmann S, Borkhardt A (2004) High expression of precursor microRNA-155/BIC RNA in children with Burkitt lymphoma. Genes Chromosomes Cancer 39:167–169

Mourelatos Z, Dostie J, Paushkin S, Sharma A, Charroux B, Abel L, Rappsilber J, Mann M, Dreyfuss G (2002) miRNPs: a novel class of ribonucleoproteins containing numerous microRNAs. Genes Dev 16:720–728

Nishikura K (2001) A short primer on RNAi: RNA-directed RNA polymerase acts as a key catalyst. Cell 107:415–418

Nykänen A, Haley B, Zamore PD (2001) ATP requirements and small interfering RNA structure in the RNA interference pathway. Cell 107:309–321

Okamura K, Siomi H, Siomi MC (2004) Distinct roles for Argonaute proteins in small RNA-directed RNA cleavage pathways. Genes Dev 18:1655–1666

Park Y, Kelley RL, Oh H, Kuroda MI, Meller VH (2002) Extent of chromatin spreading determined by roX RNA recruitment of MSL proteins. Science 298:1620–1623

Park Y, Mengus G, Bai X, Kageyama Y, Meller VH, Becker PB, Kuroda MI (2003) Sequence-specific targeting of *Drosophila* roX genes by the MSL dosage compensation complex. Mol Cell 11:977–986

Pfeffer S, Zavolan M, Grasser FA, Chien M, Russo JJ, Ju J, John B, Enright AJ, Marks D, Sander C, Tuschl T (2004) Identification of virus-encoded microRNAs. Science 304:734–736

Pham JW, Pellino JL, Lee YS, Carthew RW, Sontheimer EJ (2004) A Dicer-2-dependent 80s complex cleaves targeted mRNAs during RNAi in *Drosophila*. Cell 117:83–94

Plath K, Fang J, Mlynarczyk-Evans SK, Cao R, Worringer KA, Wang H, de la Cruz CC, Otte AP, Panning B, Zhang Y (2003) Role of histone H3 lysine 27 methylation in X inactivation. Science 300:131–135

Reik W, Walter J (2001) Genomic imprinting: parental influence on the genome. Nat Genet Rev 2:21–32

Reinhart BJ, Slack FJ, Basson M, Pasquinelli AE, Bettinger JC, Rougvie AE, Horvitz HR, Ruvkun G (2000) The 21-nucleotide let-7 RNA regulates developmental timing in Caenorhabditis elegans. Nature 403:901–906

Sado T, Wang Z, Sasaki H, Li E (2001) Regulation of imprinted X-chromosome inactivation in mice by Tsix. Development 128:1275–1286

Schaeffer C, Bardoni B, Mandel JL, Ehresmann B, Ehresmann C, Moine H (2001) The fragile X mental retardation protein binds specifically to its mRNA via a purine quartet motif. EMBO J 20:4803–4813

Schwarz DS, Hutvagner G, Du T, Xu Z, Aronin N, Zamore PD (2003) Asymmetry in the assembly of the RNAi enzyme complex. Cell 115:199–208

Seitz H, Youngson N, Lin SP, Dalbert S, Paulsen M, Bachellerie JP, Ferguson-Smith AC, Cavaille J (2003) Imprinted microRNA genes transcribed antisense to a reciprocally imprinted retrotransposon-like gene. Nat Genet 34:261–262

Seitz H, Royo H, Bortolin ML, Lin SP, Ferguson-Smith AC, Cavaille J (2004) A large imprinted microRNA gene cluster at the mouse Dlk1-Gtl2 domain. Genome Res 14:1741–1748

Sempere LF, Freemantle S, Pitha-Rowe I, Moss E, Dmitrovsky E, Ambros V (2004) Expression profiling of mammalian microRNAs uncovers a subset of brain-expressed microRNAs with possible roles in murine and human neuronal differentiation. Genome Biol 5:R13

Shibata S, Lee JT (2003) Characterization and quantitation of differential Tsix transcripts: implications for Tsix function. Hum Mol Genet 12:125–136

Slack FJ, Basson M, Liu Z, Ambros V, Horvitz HR, Ruvkun G (2000) The lin-41 RBCC gene acts in the *Caenorhabditis* heterochronic pathway between the let-7 regulatory RNA and the LIN-29 transcription factor. Mol Cell 5:659–669

Sleutels F, Zwart R, Barlow DP (2002) The non-coding Air RNA is required for silencing autosomal imprinted genes. Nature 415:810–813

Smith ER, Pannuti A, Gu W, Steurnagel A, Cook RG, Allis CD, Lucchesi JC (2000) Mol Cell Biol 20:312–318

Stavropoulos N, Lu N, Lee JT (2001) A functional role for Tsix transcription in blocking Xist RNA accumulation but not in X-chromosome choice. Proc Natl Acad Sci USA 98:10 232–10 237

Stuckenholz C, Meller VH, Kuroda MI (2003) Functional redundancy within roX1, a non-coding RNA involved in dosage compensation in *Drosophila* melanogaster. Genetics 164:1003–1014

Svoboda P, Stein P, Filipowicz W, Schultz RM (2004) Lack of homologous sequence-specific DNA methylation in response to stable dsRNA expression in mouse oocytes. Nucleic Acids Res 32:3601–3606

Tabara H, Yigit E, Siomi H, Mello CC (2002) The dsRNA binding protein RDE-4 interacts with RDE-1, DCR-1, and a DExH-box helicase to direct RNAi in *Caenorhabditis*. Cell 109:861–871

Takagi N, Sasaki M (1975) Preferential inactivation of the paternally derived X chromosome in the extraembryonic membranes of the mouse. Nature 256:640–642

Tufarelli C, Stanley JA, Garrick D, Sharpe JA, Ayyub H, Wood WG, Higgs DR (2003) Transcription of antisense RNA leading to gene silencing and methylation as a novel cause of human genetic disease. Nat Genet 34:157–165

Turner BM (2002) Cellular memory and the histone code. Cell 111:285–291

Volpe TA, Kidner C, Hall IM, Teng G, Grewal SI, Martienssen RA (2002) Regulation of heterochromatic silencing and histone H3 lysine-9 methylation by RNAi. Science 297:1833–1837

Walter J, Paulsen M (2003) Imprinting and disease. Semin Cell Dev Biol 14:101–110

Wutz A (2003) RNAs templating chromatin structure for dosage compensation in animals. Bioassays 25:434–442

Wutz A, Jaenisch R (2000) A shift from reversible to irreversible X inactivation is triggered during ES cell differentiation. Mol Cell 5:695–705

Wutz A, Rasmussen TP, Jaenisch R (2002) Chromosomal silencing and localization are mediated by different domains of Xist RNA. Nat Genet 30(2):167–174

Yekta S, Shih IH, Bartel DP (2004) MicroRNA-directed cleavage of HOXB8 mRNA. Science 304:594–596

Yi R, Qin Y, Macara IG, Cullen BR (2003) Exportin-5 mediates the nuclear export of premicroRNAs and short hairpin RNAs. Genes Dev 17:3011–3016

Zeng Y, Yi R, Cullen BR (2003) MicroRNAs and small interfering RNAs can inhibit mRNA expression by similar mechanisms. Proc Natl Acad Sci USA 100:9779–9784

Nucleic Acids and Molecular Biology, Vol. 17
Wolfgang Nellen, Christian Hammann (Eds.)
Small RNAs
© Springer-Verlag Berlin Heidelberg 2005

Potentials of a Ribozyme-Based Gene Discovery System

Masayuki Sano[1] · Kazunari Taira[1,2] (✉)

[1]Gene Function Research Center, National Institute of Advanced Industrial Science
and Technology (AIST), Central 4, 1-1-1 Higashi, 305-8562 Tsukuba Science City, Japan
taira@chembio.t.u-tokyo.ac.jp

[2]Department of Chemistry and Biotechnology, School of Engineering,
The University of Tokyo, 7-3-1 Hongo, 113-8656 Tokyo, Japan
taira@chembio.t.u-tokyo.ac.jp

Abstract The recent advent of genomic tools has provided us with very powerful ways for the identification of functional genes, thus enhancing our understanding of the molecular basis of both normal and disease phenotypes. Now that sequence information is available for many genomes, a simpler and more definitive technology for the rapid identification of functional genes is a current focus of interest. A simple screening system based on the catalytic activity of ribozymes, whose target specificities are coupled with loss-of-function mutants, has been developed to isolate key genes involved in a defined phenotype. The system was validated for functional gene screens including apoptosis, transformation, metastasis, and muscle and normal differentiation. This system should be applicable to the identification of functional genes involved in many cellular processes and diseases.

1
Introduction

Now that abundant sequence information of the human genome is readily accessible through databases, it is of great importance too that functional human genes involved in various diseases are identified and their functions as potential drug targets validated (Lander et al. 2001; Venter et al. 2001). Several powerful technologies including DNA microarrays and the yeast two-hybrid system have been developed and are widely used in identifying functional genes (Serebriiskii et al. 2001; van Berkum and Holstege 2001). The DNA microarray technique is based on the observation of changes in gene expression during particular cellular processes. However, with a DNA microarray, it seems difficult to identify genes associated with posttranslational modifications, such as phosphorylation and acetylation that cause phenotypic alterations without any change in the level of messenger RNA (mRNA). Moreover, it dose not allow a discrimination of genes that are directly responsible for a given phenotype from many other genes that are indirectly involved. The yeast two-hybrid system can detect direct interactions between proteins in vivo, and so allows the identification of genes whose products interact with target proteins derived from a complementary DNA (cDNA) library. However, in some cases, the resultant interactions do not reflect a role in the phenomenon of interest and false-positives often result. Thus, although these techniques are both sensitive and powerful, in many cases the functional involvement of isolated genes in phenotypes of interest has to be validated by additional protocols, the result being that the isolation of pseudo-positives that have to be filtered through additional protocols makes these tools seem rather tedious. Therefore, effective systems for the rapid identification of functional genes involved in a particular phenotype are clearly required. Recently, the Wong-Staal, Barber and Taira laboratories developed a novel gene discovery system using randomized ribozyme libraries (Taira et al. 1999; Kruger et al. 2000, 2001; Li et al. 2000; Welch et al. 2000; Beger et al. 2001; Chatterton et al. 2003; Nelson et al. 2003). With use of this technology, several genes related to specific phenotypes were rapidly identified (Kawasaki et al. 2002; Suyama et al. 2003a, b, 2004; Kuwabara et al. 2004; Onuki et al. 2004; Wadhwa et al. 2004; Waninger et al. 2004; Akashi et al. 2005; Sano et al. 2005). Thus, this gene discovery system should provide a sensitive and reliable tool for the rapid identification of functional genes in the postgenome era.

2
Randomized Ribozyme Libraries for Gene Discovery

Ribozymes are small versatile RNA molecules with catalytic activity; *trans*-acting ribozymes, such as the hammerhead and hairpin ribozymes, can catalyze cleavage of other RNAs at sequence-specific sites and have been extensively employed in the gene "knockdown" approach (Sarver et al. 1990; Rossi 1995; Vaish and Eckstein 1998; Shippy et al. 1999; Fedor 2000; Takagi et al. 2001; Shiota et al. 2004). The hammerhead ribozyme can cleave any RNA substrate that contains an NUH triplet, where N is any nucleotide and H is A, C or U (Fig. 1a). The hairpin ribozyme requires a GUC triplet within the target mRNA for the efficient cleavage (Fig. 1b). These ribozymes recognize and bind to their RNA targets by Watson–Crick base-paring and then cleave their targets at specific sites. Thus, ribozymes show considerable sequence flexibility in their target sequence, with the exception of the indispensable NUH or GUC triplets. To date, many ribozymes designed against viral- and disease-related genes have been tested for their ability to combat human diseases. Numerous studies have been directed towards applications of ribozymes in vivo, and many successful experiments suppressing genes of interest have been reported (Eckstein and Lilley 1996; Turner 1997; Krupp and Gaur 2000; Sioud 2004). Several ribozymes have been evaluated in clinical trials (Usman and Blatt 2000; Sullenger and Gilboa 2002; Kurreck 2003).

The use of ribozymes with randomized target recognition sequences allowed us to create a library of ribozymes capable of potentially cleaving any

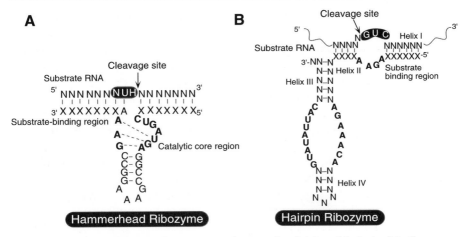

Fig. 1 Secondary structures of *trans*-acting hammerhead (**a**) and hairpin (**b**) ribozymes. The polyX corresponds to binding sites of the ribozymes

target mRNA that contains the triplet sequence recognized by ribozymes. In this gene discovery system, randomized ribozyme libraries are introduced into cells and then produce a phenotypic change (Fig. 2). When a particular ribozyme in the library binds and cleaves an mRNA that is responsible for the cell phenotype, the suppression of the target gene's expression by the ribozyme causes a phenotypic change. The ribozymes can be recovered from the cells, and subsequent sequence analysis by searching the target site sequences against gene database entries by the BLAST search program allows the identification of the targets of the ribozyme (Kruger et al. 2000, 2001; Li et al. 2000; Welch et al. 2000; Beger et al. 2001; Kawasaki et al. 2002; Suyama et al. 2003a, b, 2004; Onuki et al. 2004; Wadhwa et al. 2004; Waninger et al. 2004). Alternatively, a partial cDNA fragment that subsequently facilitates the cloning of full-length cDNA can be obtained by using 5′ and 3′ rapid amplification of cDNA ends (RACE), a PCR-based cloning technique (Kruger et al. 2000, 2001; Li et al. 2000; Welch et al. 2000; Beger et al. 2001). This system offers a method that is relatively simple and direct, compared with other high-throughput gene screening methods, and no prior sequence information is required.

Critical steps involved in this screening system include the design of the randomized library and the selection of a loss of function phenotype (Kruger et al. 2000, 2001; Li et al. 2000; Welch et al. 2000; Beger et al. 2001; Kawasaki et al. 2002; Suyama et al. 2003a, b, 2004; Onuki et al. 2004; Wadhwa et al. 2004; Waninger et al. 2004). Because the system allows the identification of direct effectors, it reduces the level of background and false-positives and produces mostly interpretable results.

The design of the randomized ribozyme library is a critical factor in this screening system. To ensure a high level of expression of both hairpin and

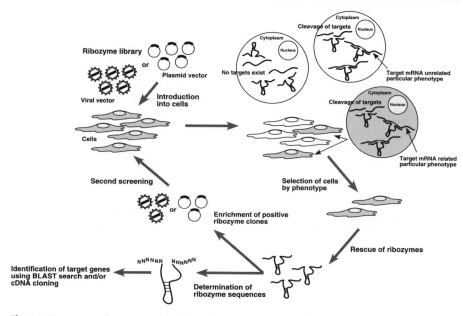

Fig. 2 Discovery of genes involved in the particular phenotype using a randomized ribozyme library. The use of the randomized ribozyme library as a tool for gene discovery involves the introduction of the library into cells, selecting cells that have a change in phenotype, isolating cells that harbored positive ribozymes, recovering ribozymes from isolated cells, reintroducing positive ribozymes into fresh cells and performing assay again (if needed), sequencing positive ribozymes and identifying the target genes involved in the particular phenotype

hammerhead ribozymes, the transfer RNA^{Val} promoter has been used successfully (Ojwang et al. 1992; Koseki et al. 1999; Kuwabara et al. 1998; Kato et al. 2001; Sano et al. 2002). The expression cassette for the ribozyme library needs to be introduced into cells using a retroviral or plasmid vector, depending on the phenotypic change to be examined. Once the assay has been performed, the effective ribozymes must be recovered to identify the complementary target mRNAs. The candidates should be validated by other ribozymes capable of cleaving the target mRNA at other sites. Small interfering RNAs (siRNAs) can also be used to knock down the gene expression and validate its involvement in the phenotype of interest (Elbashir et al. 2001; Miyagishi et al. 2002; Dykxhoorn et al. 2003; Kawasaki et al. 2003; Sioud 2004; Suyama et al. 2004; Wadhwa et al. 2004; Waninger et al. 2004). Once the novel target genes have been identified, they can be individually characterized with respect to their specific involvement in the phenotype of interest. The process can then be refined and repeated.

3
Identification of Genes Involved in Hepatitis C Virus Internal Ribosome Entry Site Mediated Translation

Hepatitis C virus (HCV) is one of the major causes of chronic hepatitis and hepatocellular carcinoma (Bradley 2000). Translation of the viral genome is mediated by an internal ribosome entry site (IRES) located in the 5′ untranslated region (UTR), and several cellular factors have been shown to bind to the 5′-UTR of HCV (Ali and Siddiqui 1995, 1997; Fukushi et al. 1997). However, the functional relevance of these factors and the potential role of additional cellular factors responsible for HCV IRES-mediated translation have not been characterized.

To identify the genes involved in HCV IRES-mediated translation, Kruger et al. (2000) made a sophisticated reporter system. An established reporter HeLa cell line contained a bicistronic reporter gene including the SV40 promoter-driving gene for hygromycin B phosphotransferase and the HCV 5′-UTR allowing for IRES-mediated translation of the herpes simplex virus thymidine kinase (HSV-tk) gene. The treatment of reporter cells with ganciclovir (GCV) induces cell death owing to the accumulation of a toxic derivative by the HSV-tk. Thus, after introducing the randomized hairpin ribozyme library by a retroviral vector, cells harboring active ribozymes that target genes involved in HCV IRES-dependent translation will survive in GCV selection. The ribozymes in the resistant cells will be recovered and then used for additional rounds of selection. Indeed, after four rounds of selection, nine dominant ribozymes were isolated, and two of these contained partially matched sequences with human eIF2γ and eIF2Bγ, respectively. Further validation indicated that ribozymes targeted against eIF2γ and eIF2Bγ could inhibit IRES-dependent translation of HCV without affecting the cap-dependent translation or cell growth.

4
Identification of Genes Involved in Cell Transformation

Using a hairpin ribozyme library, Welch et al. (2000) identified and cloned the human homologue of the *Drosophila* gene *ppan* (the Peter Pan gene) as an important factor in suppressing anchorage-independent growth of HF cells, a nontransformed revertant of HeLa cells. When HF cells were transduced with the ribozyme library using a retroviral vector, active ribozymes enhanced anchorage-independent cell growth in soft agar (Fig. 3). Repeated cycles of introducing the library led to a dramatic enrichment of the cell growth in soft agar, whereas the HF parental and the HF-control cells showed only modest enrichment. Thus, they identified a putative tumor suppressor gene, human *PPAN*, that is important in regulating anchorage dependence.

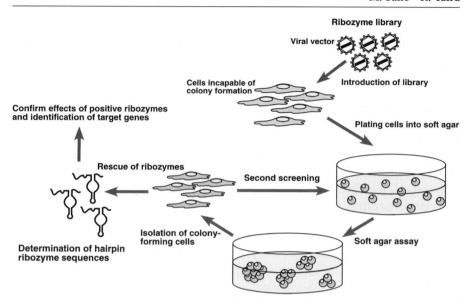

Fig. 3 A system for the identification of genes involved in anchorage-independent growth using a randomized hairpin ribozyme library. The library is introduced into cells incapable of colony formations, and then the cells are assayed for gaining the transformed phenotype in soft agar. Ribozymes are recovered from positive cells, and genes that can suppress anchorage-independent growth are identified

Interestingly, when human PPAN was exogenously expressed in HeLa cells, the cell growth in soft agar was markedly reduced compared with that for empty vector controls. This result indicated that the additional PPAN induced toxicity in HeLa cells. Thus, this example shows how relatively small differences in the expression of one gene can affect a phenotype such as anchorage-independent cell growth.

As well, Li et al. (2000) identified genes involved in cell transformation of mouse NIH3T3 cells. Although normal fibroblasts, such as NIH3T3 cells, grow as monolayers until the cells reach confluence, some cells transduced with the ribozyme library lost contact inhibition and continued to grow even after the culture cells reached confluence. The ribozymes in the transformed cells were recovered from the second round pooled foci and then sequenced. Searching for genes that contain potential matches with the recovered ribozymes through the public sequence database identified the murine telomerase reverse transcriptase (mTERT), the catalytic subunit of telomerase, as a potential candidate. This result suggested that the widely accepted role of telomerase in maintaining cellular immortalization might be more complicated than previously thought. In fact, although telomerase activity is believed to contribute to the immortalization and transformation of human cells (Hahn et al. 1999), mouse telomerase may be functionally distinct from its human counterpart (Blasco et al. 1997). Their observations demonstrated a novel role of telomerase in suppressing fibroblast transformation.

5
Identification of Genes Involved in the Pathway of Apoptosis

Cell death induced by various apoptotic stimuli, including the Fas ligand and tumor necrosis factor-α (TNF-α), occurs through signaling pathways that lead to the recruitment and activation of caspases. The Fas protein is a member of the TNFR family of TNF receptors, and TNF-α functions as a cytotoxin through binding to its receptors, TNFR-1 and TNFR-2, on the cell surface. Although Fas or TNF-α induced apoptosis has been the subject of considerable study, the details of the signaling pathways are not fully understood. Using the hammerhead ribozyme library, our group identified genes involved in both Fas-mediated and TNF-α mediated apoptosis pathways. In these studies, we employed the hybrid hammerhead ribozyme library to improve the accessibility of ribozymes to their targets; the activities of ribozymes depend on their ability to access the cleavage sites in the target mRNA in vivo (Warashina et al. 2001; Kawasaki et al. 2002). To solve the target accessibility problem, the hybrid ribozyme was generated by combining the cleavage activity of the ribozyme with the unwinding activity of the endogenous RNA helicase eIF4AI. To connect the helicase to the ribozyme, a naturally occurring RNA motif, a polyA sequence was attached to the $3'$ end of the ribozyme. This polyA sequence can interact with eIF4AI via interactions with polyA-binding protein (PABP) and PABP-interacting protein-1 (Pause and Sonenberg 1992; Craig et al. 1998).

In the Fas-mediated apoptosis study, HeLa-Fas cells were transduced by a retroviral vector that carried the randomized hybrid ribozyme library and were then treated with Fas-specific antibodies (Fig. 4). Cells harboring the hybrid ribozymes targeted against Fas-induced apoptosis-related genes could survive and the respective genomic DNA was isolated. Sequencing of the randomized region of the hybrid ribozyme in the isolated genomic DNA using the BLAST search program enabled us to rapidly identify genes involved in the apoptotic pathway. Four genes known to be involved in this process, Fas-associated death domain (FADD) protein and Caspases 3, 8 and 9, as well as additional novel genes were identified.

In the TNF-α mediated apoptosis study, TNF receptor type I associated death domain (TRADD) protein, caspases 2 and 8, Fas/TNF-α related receptor interacting protein (RIP), RIP-associated ICH-1/CED-3-homologous protein with death domain (RAIDD), TNF receptor associated factor 2 (TRAF 2), Ets 1 and Bak were identified when MCF-7 cells harboring the randomized hybrid ribozyme library were treated with TNF-α (Kawasaki et al. 2002). In addition to the aforementioned genes that are known to have apoptotic func-

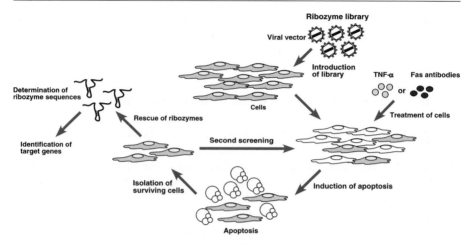

Fig. 4 A system for the identification of genes involved in apoptosis. A hammerhead ribozyme library is used to identify genes involved in Fas-mediated or tumor necrosis factor-α (*TNF-α*) mediated apoptosis. After introducing the library into cells, apoptosis is induced by Fas-specific antibodies or TNF-α treatment

tions during TNF-α induced apoptosis, partial sequences for 30 new target genes were also isolated. Further investigation of the genes identified in these studies would help to revealing the complex pathways of apoptosis.

Importantly, the addition of a polyA tail to the ribozymes reduced the number of false-positives in both studies. Indeed, with the use of conventional randomized ribozymes without the polyA tail, FADD and caspase 8 were not identified in the Fas-mediated apoptosis study, demonstrating the integrity of the hybrid system. Wadhwa et al. (2003) also demonstrated that hybrid ribozymes are far more effective than their conventional counterparts in knocking down the expression of an hsp70 family protein, mortalin. Whereas conventional ribozymes were ineffective, hybrid ribozymes suppressed mortalin expression and resulted in growth suppression of human transformed cells.

6
Identification of Metastasis-Related Genes

Metastasis is often one of the most serious problems in cancer therapy. Investigations into the mechanisms of metastasis using DNA microarrays have revealed that the motility of several strong metastatic cancer cell lines is greater than that of nonmetastatic or weakly metastatic cell lines, and that motility is a prerequisite for metastasis (Clark et al. 2000; Hippo et al. 2001). However, the underlying mechanisms remain unclear because metastasis is a complex phenomenon that involves many genes and pathways. To clarify the mechanism of metastasis, Suyama et al. (2003a) from our laboratory have focused on identifying genes involved in cell migration. A highly invasive derivative clone of HT1080 fibrosarcoma cells was transfected with a plasmid vector harboring a conventional randomized hammerhead ribozyme library and subjected to a chemotaxis assay in 12-μm-pore Transwell inserts with fibronectin as a chemoattractant (Fig. 5). Although HT1080 cells normally migrate toward a chemoattractant (Varani 1982), some cells did not migrate toward the chemoattractant after treatment with the randomized ribozyme library. Sequencing of the randomized regions of the ribozymes isolated from the cells that remained in the top chamber and a search for the relevant sequences in databases allowed for the identification of genes that are likely to play a role in metastasis. Two ribozymes directed against the transcript of *ROCK-1*, one of the genes that regulate the actin cytoskeleton, were found and these ribozymes inhibited the migration of HT1080 cells almost completely without any effect on their proliferation. The Rho–ROCK system has previously been implicated in cell mobility, migration and metastasis for a variety of cell types (Itoh et al. 1999). Therefore, ribozymes targeted against *ROCK-1* may lead to the inhibition of cell migration in several cancer cell lines, and those ribozymes and their alternative inhibitors targeted to *ROCK-1* mRNA might be useful in cancer therapy as low-toxicity compounds inhibiting cell motility without affecting cell viability.

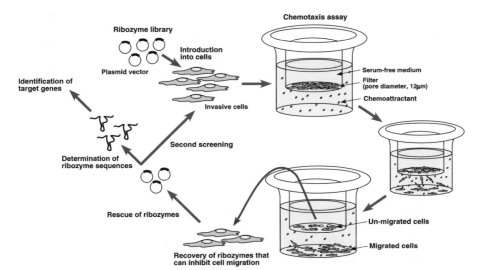

Fig. 5 A system for the identification of genes involved in metastasis. A hammerhead ribozyme library is used to identify genes that promote cell migration. After introducing it into invasive cells, chemotaxis assay is performed using Transwells with a chemoattractant. Cells with reduced migration are isolated from the top chamber, and positive ribozymes are rescued and sequenced to identify genes that can promote cell migration

Suyama et al. (2003b) also identified genes involved in enhancing the invasive properties of noninvasive NIH3T3 cells in an extracellular matrix (ECM) coated Transwell using a hybrid ribozyme library (Suyama et al. 2003b). The Transwell with ECM gel serves as a barrier to block noninvasive cells from migrating through the filter (Albini et al. 1987). Cells transduced with the hybrid ribozyme library were subjected to invasion assay and cells that had migrated to the lower compartment through the ECM gel were collected. Genes identified by this assay included Gem GTPase, and uncharacterized genes that resemble myosine phosphatase and protein tyrosine phosphatase.

A randomized hammerhead ribozyme has also been used in in vivo metastasis studies by injecting C57BL/6N mice with mouse B16F0 low-invasive melanoma cells harboring the ribozyme library (Suyama et al. 2004). Two weeks after injection, the number of metastatic melanoma nodules derived from library-treated cells was higher than that derived from untreated cells. Cells were collected from the pulmonary nodules of melanoma cells derived from ribozyme harboring cells, and the ribozymes were then amplified by a reverse transcriptase PCR strategy and cloned. Sequencing of the ribozymes identified eight types of ribozymes that contained specific substrate binding arms. One of the target genes identified was the gene for stromal interaction molecule 1 (STIM1), which is a gene for a transmembrane glycoprotein involved in tumorigenesis. The database search also identified genes whose functions have not yet been well characterized, such as the sequence AW551984, expressed in mouse, which contains a VWA domain that seems to mediate adhesion of eukaryotic cells. The roles of these genes in cell migration and invasion are currently being investigated to further elucidate their functions in pathways that promote cancer metastasis. Further investigation of genes identified would provide important clues to the complex mechanism(s) of metastasis.

7
Identification of Genes Involved in Alzheimer's Disease

Endoplasmic reticulum (ER) stress seems to be associated with certain genetic and/or neuronal degenerative diseases, such as Alzheimer's disease (AD). Several possible pathogenic mechanisms have been proposed for AD, including mutations in or expression of alternatively spliced variants of presenilin, impairment of the signaling of ER stress and increased sensitivity to stress-induced apoptosis (Katayama et al. 1999, 2001; Sato et al. 2001).

To identify genes involved in AD, Onuki et al. (2004) from our group focused on genes that are involved in tunicamycin-induced ER stress-mediated apoptosis. When human neuroblastoma SK-N-SH cells harboring the randomized hammerhead ribozyme library were treated with tunicamycin, the gene for a double-stranded RNA (dsRNA) dependent protein kinase (PKR) was isolated from a surviving cell. In tunicamycin-treated SK-N-SH cells, further analyzes revealed that levels of both PKR and phosphorylated PKR were elevated in the nuclei, where PKR formed aggregates. In addition, the expression of mutated PKR that cannot take an active (phosphorylated) form remained in the cytoplasm and inhibited apoptosis after the treatment of cells with tunicamycin. Furthermore, this study also revealed that PKR localized in the nuclei of autopsied hippocampus of AD patients, and that the level of phosphorylated PKR in nuclear fractions from AD patients was higher than in that from disease-free controls. These results suggested that phosphorylation and aggregation of PKR in the nucleus of neurons could be associated with the pathology of AD.

8
Conclusions

The randomized ribozyme library is a very powerful tool for identifying genes involved in specific pathways and phenotypes. In addition, this system can be used to study other aspects of basic molecular and cellular biology, as well as the pathology of diseases, such as from differentiation, proliferation, aging and neurodegeneration (Rhoades and Wong-Staal 2003). Recently, Kuwabara et al. (2004) identified a noncoding dsRNA, namely, small modulatory dsRNA, which plays a critical role in mediating neuronal differentiation. Small noncoding RNAs including microRNAs and siRNAs have emerged as crucial components in the regulation of many cellular phenomena such as differentiation, cell proliferation, cell death and cell metabolism (Bartel 2004). As structural and sequence information of these noncoding small RNAs accumulates, the gene discovery system using ribozyme libraries could become a valuable tool for the identification of many genes for small RNAs that involve particular phenotypes.

Acknowledgement The authors would like to thank Renu Wadhwa for comments, suggestions and helpful discussion.

References

Akashi H, Matsumoto S, Taira K (2005) Gene discovery by ribozyme and siRNA libraries. Nat Rev Mol Cell Biol 6:413–422

Albini A, Iwamoto Y, Kleinman HK, Martin GR, Aaronson SA, Kozlowski JM, McEwan RN (1987) A rapid in vitro assay for quantitating the invasive potential of tumor cells. Cancer Res 47:3239–3245

Ali N, Siddiqui A (1995) Interaction of polypyrimidine tract-binding protein with the 5' noncoding region of the hepatitis C virus RNA genome and its functional requirement in internal initiation of translation. J Virol 69:6367–6375

Ali N, Siddiqui A (1997) The La antigen binds 5' noncoding region of the hepatitis C virus RNA in the context of the initiator AUG codon and stimulates internal ribosome entry site-mediated translation. Proc Natl Acad Sci USA 94:2249–2254

Bartel DP (2004) MicroRNAs: genomics, biogenesis, mechanism, and function. Cell 116:281–297

Beger C, Pierce LN, Kruger M, Marcusson EG, Robbins JM, Welcsh P, Welch PJ, Welte K, King MC, Barber JR, Wong-Staal F (2001) Identification of Id4 as a regulator of BRCA1 expression by using a ribozyme-library-based inverse genomics approach. Proc Natl Acad Sci USA 98:130–135

Blasco MA, Lee HW, Hande MP, Samper E, Lansdorp PM, DePinho Greider CW (1997) Telomere shortening and tumor formation by mouse cells lacking telomerase RNA. Cell 91:25–34

Bradley DW (2000) Studies of non-A, non-B hepatitis and characterization of the hepatitis C virus in chimpanzees. Curr Top Microbiol Immunol 242:1–23

Chatterton JE, Hu X, Wong-Staal (2004) Ribozymes in gene identification, target validation, and drug discovery. TARGETS 3:10–17

Clark EA, Golub TR, Lander ES, Hynes RO (2000) Genomic analysis of metastasis reveals an essential role for RhoC. Nature 406:532–535

Craig AW, Haghighat A, Yu AT, Sonenberg N (1998) Interaction of polyadenylate-binding protein with the eIF4G homologue PAIP enhances translation. Nature 392:520–523

Dykxhoorn DM, Novina CD, Sharp PA (2003) Killing the messenger: short RNAs that silence gene expression. Nat Rev Mol Cell Biol 4:457–467

Eckstein F, Lilley DMJ (eds) (1996) Nucleic acids and molecular biology: catalytic RNA, vol 10. Springer, Berlin Heidelberg New York

Elbashir SM, Harborth J, Lendeckel W, Yalcin A, Weber K, Tuschl T (2001) Duplexes of 21-nucleotide RNAs mediate RNA interference in cultured mammalian cells. Nature 411:494–498

Fedor MJ (2000) Structure and function of the hairpin ribozyme. J Mol Biol 297:269–291

Fukushi S, Kurihara C, Ishiyama N, Hoshino FB, Oya A, Katayama K (1997) The sequence element of the internal ribosome entry site and a 25-kilodalton cellular protein contribute to efficient internal initiation of translation of hepatitis C virus RNA. J Virol 71:1662–1666

Hahn WC, Counter CM, Lundberg AS, Beijersbergen RL, Brooks MW, Weinberg RA (1999) Creation of human tumour cells with defined genetic elements. Nature 400:464–468

Hippo Y, Yashiro M, Ishii M, Taniguchi H, Tsutsumi S, Hirakawa K, Kodama T, Aburatani H (2001) Differential gene expression profiles of scirrhous gastric cancer cells with high metastatic potential to peritoneum or lymph nodes. Cancer Res 61:889–895

Itoh K, Yoshioka K, Akedo H, Uehata M, Ishizaki T, Narumiya S (1999) An essential part for Rho-associated kinase in the transcellular invasion of tumor cells. Nat Med 5:221–225

Katayama T, Imaizumi K, Sato N, Miyoshi K, Kudo T, Hitomi J, Morihara T, Yoneda T, Gomi F, Mori Y et al (1999) Presenilin-1 mutations downregulate the signalling pathway of the unfolded-protein response. Nat Cell Biol 1:479–485

Katayama T, Imaizumi K, Honda A, Yoneda T, Kudo T, Takeda M, Mori K, Rozmahel R, Fraser P, George-Hyslop PS, Tohyama M (2001) Disturbed activation of endoplasmic reticulum stress transducers by familial Alzheimer's disease-linked presenilin-1 mutations. J Biol Chem 276:43446–43454

Kato Y, Kuwabara T, Warashina M, Toda H, Taira K (2001) Relationships between the activities in vitro and in vivo of various kinds of ribozyme and their intracellular localization in mammalian cells. J Biol Chem 276:15378–15385

Kawasaki H, Onuki R, Suyama E, Taira K (2002) Identification of genes that function in the TNF-alpha-mediated apoptotic pathway using randomized hybrid ribozyme libraries. Nat Biotechnol 20:376–380

Koseki S, Tanabe T, Tani K, Asano S, Shioda T, Nagai Y, Shimada T, Ohkawa J, Taira K (1999) Factors governing the activity in vivo of ribozymes transcribed by RNA polymerase III. J Virol 73:1868–1877

Kruger M, Beger C, Li QX, Welch PJ, Tritz R, Leavitt M, Barber JR, Wong-Staal F (2000) Identification of eIF2Bgamma and eIF2gamma as cofactors of hepatitis C virus internal ribosome entry site-mediated translation using a functional genomics approach. Proc Natl Acad Sci USA 97:8566–8571

Kruger M, Beger C, Welch PJ, Barber JR, Manns MP, Wong-Staal F (2001) Involvement of proteasome alpha-subunit PSMA7 in hepatitis C virus internal ribosome entry site-mediated translation. Mol Cell Biol 21:8357–8364

Krupp G, Gaur RK (eds) (2000) Ribozyme, Biochemistry and biotechnology. Eaton, Natick, MA

Kurreck J (2003) Antisense technologies. Improvement through novel chemical modifications. Eur J Biochem 270:1628–1644

Kuwabara T, Warashina M, Tanabe T, Tani K, Asano S, Taira K (1998) A novel allosterically trans-activated ribozyme, the maxizyme, with exceptional specificity in vitro and in vivo. Mol Cell 2:617–627

Kuwabara T, Hsieh J, Nakashima K, Taira K, Gage FH (2004) A small modulatory dsRNA specifies the fate of adult neural stem cells. Cell 116:779–793

Lander ES, Linton LM, Birren B, Nusbaum C, Zody MC, Baldwin J, Devon K, Dewar K, Doyle M, FitzHugh W et al. (2001) Initial sequencing and analysis of the human genome. Nature 409:860–921

Li QX, Robbins JM, Welch PJ, Wong-Staal F, Barber JR (2000) A novel functional genomics approach identifies mTERT as a suppressor of fibroblast transformation. Nucleic Acids Res 28:2605–2612

Miyagishi M, Taira K (2002) U6 promoter-driven siRNAs with four uridine 3′ overhangs efficiently suppress targeted gene expression in mammalian cells. Nat Biotechnol 20:497–500

Nelson DL, Suyama E, Kawasaki H, Taira K (2003) Use of random ribozyme libraries for the rapid screening of apoptosis- and metastasis-related genes. TARGETS 2:191–200

Ojwang JO, Hampel A, Looney DJ, Wong-Staal F, Rappaport J (1992) Inhibition of human immunodeficiency virus type 1 expression by a hairpin ribozyme. Proc Natl Acad Sci USA 89:10 802–10 806

Onuki R, Bando Y, Suyama E, Katayama T, Kawasaki H, Baba T, Tohyama M, Taira K (2004) An RNA-dependent protein kinase is involved in tunicamycin-induced apoptosis and Alzheimer's disease. EMBO J 23:959–968

Pause A, Sonenberg N (1992) Mutational analysis of a DEAD box RNA helicase: the mammalian translation initiation factor eIF-4A. EMBO J 11:2643–2654

Rhoades K, Wong-Staal F (2003) Inverse genomics as a powerful tool to identify novel targets for the treatment of neurodegenerative diseases. Mech Ageing Dev 124:125–132

Rossi JJ (1995) Controlled, targeted, intracellular expression of ribozymes: progress and problems. Trends Biotechnol 13:301–306

Sano M, Kuwabara T, Warashina M, Fukamizu A, Taira K (2002) Novel method for selection of tRNA-driven ribozymes with enhanced stability in mammalian cells. Antisense Nucleic Acid Drug Dev 12:341–352

Sano M, Kato Y, Taira K (2005) Functional gene-discovery systems based on libraries of hammerhead and hairpin ribozymes and short hairpin RNAs. Mol Biosystems 1:27–35

Sarver N, Cantin EM, Chang PS, Zaia JA, Ladne PA, Stephens DA, Rossi JJ (1990) Ribozymes as potential anti-HIV-1 therapeutic agents. Science 247:1222–1225

Sato N, Imaizumi K, Manabe T, Taniguchi M, Hitomi J, Katayama T, Yoneda T, Morihara T, Yasuda Y, Takagi T et al. (2001) Increased production of beta-amyloid and vulnerability to endoplasmic reticulum stress by an aberrant spliced form of presenilin 2. J Biol Chem 276:2108–2114

Serebriiskii IG, Khazak V, Golemis EA (2001) Redefinition of the yeast two-hybrid system in dialogue with changing priorities in biological research. Biotechniques 30:634–636

Shiota M, Sano M, Miyagishi M, Taira K (2004) Ribozyme: application to functional analysis and gene discovery. J Biochem 136:133–147

Shippy R, Lockner R, Farnsworth M, Hampel A (1999) The hairpin ribozyme. Discovery, mechanism, and development for gene therapy. Mol Biotechnol 12:117–129

Sioud M (ed) (2004) Methods in molecular biology, vol 252. Ribozymes and siRNA protocols. Humana, Totowa

Sullenger BA, Gilboa E (2002) Emerging clinical applications of RNA. Nature 418:252–258

Suyama E, Kawasaki H, Kasaoka T, Taira K (2003a) Identification of genes responsible for cell migration by a library of randomized ribozymes. Cancer Res 63:119–124

Suyama E, Kawasaki H, Nakajima M, Taira K (2003b) Identification of genes involved in cell invasion by using a library of randomized hybrid ribozymes. Proc Natl Acad Sci USA 100:5616–5621

Suyama E, Wadhwa R, Kaur K, Miyagishi M, Kaul SC, Kawasaki H, Taira K (2004) Identification of metastasis-related genes in a mouse model using a library of randomized ribozymes. J Biol Chem 279:38 083–38 086

Taira K, Warashina M, Kuwabara T, Kawasaki H (1999) Functional hybrid molecules with sliding ability. Jpn Patent Appl H11–316 133

Takagi Y, Warashina M, Stec WJ, Yoshinari K, Taira K (2001) Recent advances in the elucidation of the mechanisms of action of ribozymes. Nucleic Acids Res 29:1815–1834

Turner PC (ed) (1997) Methods in molecular biology, vol 74. Ribozyme protocols. Humana, Totowa

Usman N, Blatt LM (2000) Nuclease-resistant synthetic ribozymes: developing a new class of therapeutics. J Clin Invest 106:1197–1202

Vaish NK, Kore AR, Eckstein F (1998) Recent developments in the hammerhead ribozyme field. Nucleic Acids Res 26:5237–524

van Berkum NL, Holstege FC (2001) DNA microarrays: raising the profile. Curr Opin Biotechnol 12:48–52

Varani J (1982) Chemotaxis of metastatic tumor cells. Cancer Metastasis Rev 1:17–28

Venter JC, Adams MD, Myers EW, Li PW, Mural RJ, Sutton GG, Smith HO, Yandell M, Evans CA, Holt RA et al (2001) The sequence of the human genome. Science 291:1304–1351

Wadhwa R, Ando H, Kawasaki H, Taira K, Kaul SC (2003) Targeting mortalin using conventional and RNA-helicase-coupled hammerhead ribozymes. EMBO Rep 4:595–601

Wadhwa R, Yaguchi T, Kaur K, Suyama E, Kawasaki H, Taira K, Kaul SC (2004) Use of a randomized hybrid ribozyme library for identification of genes involved in muscle differentiation. J Biol Chem 279:51 622–51629

Waninger S, Kuhen K, Hu X, Chatterton JE, Wong-Staal F, Tang H (2004) Identification of cellular cofactors for human immunodeficiency virus replication via a ribozyme-based genomics approach. J Virol 78:12829–12837

Warashina M, Kuwabara T, Kato Y, Sano M, Taira K (2001) RNA-protein hybrid ribozymes that efficiently cleave any mRNA independently of the structure of the target RNA. Proc Natl Acad Sci USA 98:5572–5577

Welch PJ, Marcusson EG, Li QX, Beger C, Kruger M, Zhou C, Leavitt M, Wong-Staal F, Barber JR (2000) Identification and validation of a gene involved in anchorage-independent cell growth control using a library of randomized hairpin ribozymes. Genomics 66:274–283

Subject Index

Aberrant RNA 120, 121, 153, 154, 162–164, 187, 189, 190
ADAR 94, 98, 99, 103, 104, 108, 109, 143
Algorithm 68, 69, 78–80, 85
Alignment 58, 59, 62–64, 67, 70, 83, 84
Antisense RNA *see* asRNA
Antiviral defence 91, 99, 102, 104, 110, 111
Argonaute proteins 99, 105, 142, 144, 146, 159, 162, 166, 168, 182–185
asRNA 1–10, 13–21, 24, 91–93, 124, 141, 142, 147, 149–155, 163–165, 179
– activator 14
– *cis*-encoded 13, 17, 19, 24, 92, 142
– coupled degradation 18
– inhibitory 14, 149
– Natural antisense transcript (NAT) 92, 93
– plasmid encoded 17–21
– *trans*-encoded 16, 17, 21, 92, 142

Base pairing
– association rate 20–21
– degree of complementarity 16, 75, 77, 80, 83–86, 92, 93, 98, 102, 104, 107–111, 145, 163, 179, 180, 182, 184
– dissociation rate 21
– stability 20, 21
Biofilm 11, 14
Bioluminescence 14

CopA/CopT 21, 93
Co-suppression 119, 124, 125, 128, 129, 150

Ded-1 147
Dicer 75, 77, 98, 99, 101, 104–108, 110, 119, 120, 122, 124, 142–154, 162, 164–166, 168–171, 180–185, 187, 188
DNA elimination 106, 161, 168
DNA methylation 109, 161, 163, 166, 168–170, 172, 188–190, 194
Double-stranded RNA *see* dsRNA
Drh-1 182
Drosha 77, 104, 106, 108, 181, 182
dsRBD 92, 96–106, 142, 143, 146,
dsRNA
– length 64, 91, 94, 107, 108
– localisation 91, 92, 94, 107, 108
– unwinding 105, 148, 183

Editing 96, 98, 103, 104, 108, 109, 143
EF-Tu 147
EGO-1 121, 123
Eri-1 134, 147–149, 155
Exportin 5 108, 181

Gulliver 160, 164, 165, 170, 171

HCV
– IRES 205
HelF 147–149, 152, 154
Heterochromatin 106, 109, 161, 163, 166–170, 172, 187, 192
Heterochronic
– genes 76
– RNA 142
Hfq 4, 8–10, 12, 15, 17–20, 23, 24
Histone methylation 109, 167–172, 186, 192–194
– H3K4 167, 171, 172
– H3K9 168, 170, 193
Hox genes 185, 186

Imprinting 43, 186, 187, 189, 190, 194
Intergenic region 2, 34, 44, 78, 84, 171

Long direct repeat 13

Microarray 4, 34, 85

Printing: Krips bv, Meppel
Binding: Stürtz, Würzburg